Revising Professional Writing

in Science and Technology, Business, and the Social Sciences

Kathryn Riley
University of Minnesota, Duluth

Kim Sydow Campbell
University of Alabama

Alan Manning
Brigham Young University

Frank Parker
Louisiana State University (Retired)

PARLAY PRESS

Superior, Wisconsin

Parlay Press
P.O. Box 894
Superior, WI 54880
Telephone/Fax (218) 834-2508

Kathryn Riley, Kim Sydow Campbell, Alan Manning, and Frank Parker

Revising Professional Writing
ISBN 0-9644636-6-0

Contents

Introduction

Revising Professional Writing replaces and expands upon *Revising Technical and Business Writing, 2nd edition*. In updating that book, we have added examples and exercises that deal not only with science and technology and business, but also with the social sciences and human services. *Revising Professional Writing* is designed as a supplementary text for such courses as technical and scientific writing, business writing, writing in the social sciences, and other writing courses beyond freshman composition. The primary textbook in such courses usually focuses on planning and drafting specific types of letters, memos, and reports. In contrast, *Revising Professional Writing* focuses on the second half of the writing process: revising and editing.

An overview of this book

Revising Professional Writing covers four areas that are relevant to all types of professional writing:

- Part I, **Development**, shows how to revise the amount and kind of detail in informative and persuasive prose and how to supplement such prose with graphics.
- Part II, **Organization**, shows how to revise the arrangement of parts of a text and how to provide signals to help the reader follow this arrangement.
- Part III, **Style and Tone**, shows how to revise sentences to achieve goals such as conciseness, appropriate use of active and passive voice, and effective word choice.
- Part IV, **Punctuation and Grammar**, shows how to edit common problems in punctuation and sentence construction, as well as how to integrate references to outside sources.

Revising Professional Writing contains a variety of materials to guide students through the revision process. First, each chapter reviews the major PRINCIPLES involved in one element of revision. The discussion of principles also includes numerous examples, often presented in "before and after" format. Second, each chapter is followed by APPLICATIONS, opportunities to revise sentences and paragraphs using the principles discussed in the chapter. Most of these applications are based on texts actually written by college students in upper-division writing classes. Third, the book includes LONGER TEXTS FOR REVISION. These texts, which build upon the shorter applications, call for more complex revision tasks in the context of longer pieces of writing.

General guidelines for the revision process

The following general guidelines reflect the revision strategies followed by many professional writers, as well as suggestions that many student writers have found helpful.

Let your draft sit for a while. Build enough time into your writing schedule to let your early draft sit for at least a few hours or, preferably, overnight. You'll find it easier to look at your draft objectively after you have some distance on it.

Word-process your draft. Even if you begin with a handwritten version, construct a word-processed draft to revise and edit. A word-processed version will give you a better sense of features such as paragraph length, format, and so on. In addition, having your draft on disk makes it much easier to revise, since you can easily move, add, and delete text.

Revise one element at a time. Rather than trying to revise development, organization, style and tone, and grammar and punctuation in one pass, try going over your draft more than once, focusing on a different element each time. In the long run, this procedure is more efficient, since you are less likely to overlook problems in a particular area.

Revise "global" elements before "local" elements. Try revising development and organization before moving to sentence-level features such as style, tone, grammar, and punctuation. Otherwise, if you start by editing a feature such as punctuation and then add text, you will have to edit the new text again for punctuation.

Ask someone else to read your draft. Having a classmate or colleague read your draft can be illuminating, especially if you are writing to an audience who knows less about your subject than you do. Often, another reader can more easily spot terms that need definition, generalizations that need examples, or arguments that need more support. Many writing teachers use peer editing workshops to introduce this revision strategy.

Use a revision checklist. Create a short list of the features that your final draft should have, or the problems that you want to avoid, and use this checklist as an aid during the revision process. (If you have a chance to talk with your potential reader before beginning a writing project, you can incorporate the reader's preferences into the checklist.)

Read your draft aloud. Reading aloud passages from your draft can help you spot problems in style (e.g., overly long sentences) that may not show up during silent reading. This strategy can also you achieve a more conversational style in letters or memos.

Review examples of similar documents. As you encounter memos, letters, reports, and articles, read them critically for their properties as pieces of writing. When you run across a highly effective document, file it for future reference. Sometimes examining an effective piece of writing can help you identify what's missing in your own draft.

Chapter 1. Revising Informative Prose

One important difference between student writing and professional writing lies in how much the writer and reader are expected to know about the subject matter of a text. Most student writing is addressed to a faculty member — a reader who already knows a good deal about the subject matter. The purpose of student writing is typically to demonstrate the writer's understanding to an already knowledgeable reader.

In contrast, in your role as a professional, this rhetorical situation is usually reversed. Instead of addressing an already knowledgeable reader, you will often be writing to inform the reader: to report events, explain a process, give instructions, describe an object, analyze a situation, and so on.

Revising informative prose requires attention to its two essential elements: the CONTROLLING IDEA of the passage and the SUPPORTING DETAIL that expands upon it. The controlling idea consists of a generalization (usually stated in a verb phrase) about a subject (usually stated in a noun phrase). For example, the sentence *Businesses can be divided into three types, depending on their ownership structure* states a generalization (*can be divided into three types, depending on their ownership structure*) about a subject (*businesses*). Strategies for revising controlling ideas are discussed in Chapter 4, "Revising for Paragraph Unity." The present chapter focuses on ways to develop supporting detail: facts and explanations that lead the reader to understand and accept the controlling idea.

Defining

Technical terms should be briefly defined the first time you use them if your audience consists partly or entirely of nonspecialists. One common pattern for short definitions is the FORMAL DEFINITION, which consists of three parts: the term to be defined, the larger class that the term belongs to, and the features that distinguish the term from other members of its class. This technique is used in example (1).

(1) The term *interlanguage* refers to an interim grammar (i.e., a set of rules) that a speaker constructs while learning a second language.

Here the term being defined (*interlanguage*) is placed into the larger class of grammars; it is differentiated from other grammars by its interim nature and its association with second-language acquisition. This example also uses a SYNONYM (*set of rules*) to stipulate the sense in which grammar is being used. In addition, the term being defined is set off typographically, in this case with italics. This visual cue tells the reader that a new technical term is being introduced. (For other similar techniques, see Chapter 7, "Revising Format to Unify Text.")

Another way to define a term is to provide an OPERATIONAL DEFINITION, which describes the conditions under which the item being defined can be observed. Example (2) contains operational definitions of the terms *trash, mulch*, and *compost*.

(2) If you throw away leaves, twigs, grass clippings, or sawdust, you create trash. If you mix those materials and use them right away, you create mulch. And if you mix them,

perhaps add vegetable scraps from the kitchen, and let the stuff decompose, you create compost. ("Mulching Mowers," *Consumer Reports*, June 1991, p. 411)

You may want to provide an EXTENDED DEFINITION of one or more paragraphs. An extended definition typically begins with a brief definition and then uses other development techniques such as descriptions, examples, and comparisons. Example (3) gives an extended definition of *social stereotypes* as the term is used in the study of language and society.

(3)   Speakers may treat some features as *social stereotypes*, where they comment overtly on their use. Items such as *ain't*, double negatives, and *dese, dem,* and *dose* are classic features of this type. . . . Sociolinguistic stereotypes tend to be overly categorical and are often linguistically naive, although they may derive from a basic sociolinguistic reality. For example, the stereotype that working class speakers *always* use *dese, dem*, and *dose* forms and middle-class speakers *never* do is not supported empirically, although there certainly is a correlation between the relative frequency of the nonstandard variant and social stratification. (W. Wolfram, *Dialects and American English*, Englewood Cliffs, NJ: Prentice Hall, 1991, p. 99)

Note that where a formal definition is more like a dictionary entry, an extended definition is more like an encyclopedia entry.

Describing

Describing an object typically involves giving details about features that can be physically verified — for example, size, shape, color, weight, material, and so forth. The goal of such descriptions in technical writing is usually to convey information as objectively as possible: that is, in terms that will be interpreted the same way by reader and writer. For this reason, physical descriptions often use quantifiable terms, so that words with a range of interpretations (e.g., *large*) are limited to a particular interpretation (e.g., 18 feet). Therefore, one strategy in revising physical descriptions is to quantify terms that can be interpreted in more than one way. For example, the description of a peregrine in (4) includes details that help the reader visualize its size and speed.

(4)   The adult peregrine falcon (*Falco peregrinus*) is among the world's fastest and fiercest birds, a powerful crow-size raptor that catches dinner on the wing or kills prey by "stooping" in precise vertical dives clocked at more than 200 miles per hour. (G. Rowell, "Falcon Rescue," *National Geographic*, April 1991, p. 108)

This passage gives the reader a more precise image than simply saying that the peregrine is a "fast, fairly large bird."

In addition to descriptions of physical objects, descriptions of more abstract entities such as processes, behaviors, and programs are often required of professional writers. As with descriptions of objects, the goal is to include the amount and kind of detail that will allow the reader to understand the writer's view of what is being described. For example, (5) describes an experimental procedure used to test language perception and production in three children.

(5)   Each subject was asked to identify and produce a word pair contrasting in final position. Subject A was tested on *pants-pan* and *pat-pad*; J, on *pat-pad*; and D, on *pats-pad*. The identification task involved pointing to one of two pictures when a member of the word pair was spoken by the experimenter. Each member of the pair was presented eight times in a previously determined, randomized order. All three children scored from 94 to 100% correct on this task. . . .

   Production testing immediately followed the identification task. The experimenter asked each child to label each picture when it was pointed to. Four or five productions of each word were recorded in a randomized order. During the production task, each child wore a toy hat with a Sony ECM150 microphone attached at a stable distance (about 6 in) in front of the child's mouth. Responses were recorded using a Sony WM-D6C cassette recorder. (K. Riley, P. R. Hoffman, and S. K. Damico, "The Effects of Conflicting Cues on the Perception of Misarticulations," *Journal of Phonetics*, 1986, pp. 481-487)

Note that an experimental report, such as this one, should provide enough detail about the subjects and procedures to allow the reader to replicate the study, at least in theory.

Giving Examples

Examples are a highly effective way to develop generalizations, unfamiliar concepts, or abstract ideas, especially when writing to a nonspecialist audience. Passage (6) uses examples to illustrate an IRS rule for deductible educational expenses.

(6)   To qualify [for deduction], the education must be in subject matter related to your teaching, research, or other professional duties. For example, a German teacher would be allowed to deduct expenses for taking a course in German literature. The education in question may be day courses, vocational courses, research on a dissertation, or any other educational activity. . . .

   Education which contributes to your general education but is not directly related to your professional duties does not qualify for deduction. For example, a mathematics teacher would not ordinarily be allowed a deduction for taking a course in political science. (Allen Bernstein, *1990 Tax Guide for College Teachers*, Washington, DC: Academic Information Service, 1989, pp. 373-374)

Generally speaking, the less familiar the concept, the more examples the reader will need to understand it. In addition, the examples should be representative — that is, they should illustrate the range of items to which a term typically refers. For instance, explaining the term *parasite* by using only fleas and lice as examples might lead the reader to conclude that the term describes only bloodsucking insects. Including examples such as mistletoe and lamprey eels would prevent this mistaken conclusion.

Comparing and contrasting

Comparisons and analogies can help the reader understand the unfamiliar (by showing how it is like the familiar) or the abstract (by showing how it is like the concrete). A comparison focuses on similarities between items from the same domain — for example, two printers, two employee development programs, two experimental techniques. Excerpt (7), from a power company newsletter, compares the cost of electricity with the cost of other consumer products, based on an average cost of 5.6 cents per kilowatt-hour.

(7)   For the cost of one pound of gourmet-type coffee (about $6) you can brew one pot a day for over a year.

The cost of admission to a movie (about $4) will operate your television for 3 hours a day for 4 months.

The cost of an average magazine (about $2.50) will operate a 60-watt lightbulb by which you can read it for more than 700 hours.

The cost of a low-calorie microwave dinner (about $3) will operate a microwave oven for about 33 hours. (*Minnesota Power Energizer*, March/April 1990, p. 4)

An analogy compares items that are from different domains, but that have essential similarities. Passage (8) illustrates an abstract concept — particles of light — by drawing an analogy with the action of a shotgun.

(8)   When an intense light source (such as a high energy laser) is turned on, it undergoes a slight but measurable "kick," comparable to the "kick" of a shotgun. The kick of a shotgun results from the ejection of particles (the shot) at high velocity. This suggests that a beam of light likewise consists of particles leaving the source at high velocity.

As these examples show, comparisons and analogies are most useful if they link the object being explained to one that is more familiar or easier to visualize.

While comparisons and analogies focus on similarities, contrastive analyses focus on differences between related objects. For example, passage (9) contrasts surgical biopsy and stereotaxic fine-needle biopsy (SFNB).

(9)   SFNB has several advantages over surgical biopsy. First, because no incision is made and only a few cells are removed, it can be performed easily and without delay. Results are often available immediately or at last the same day. Second, it is relatively pain free. Biopsy, on the other hand, is considered minor surgery and requires local — and sometimes general — anesthesia. The patient usually has to wait a few anxiety-ridden weeks before it can be scheduled, and because a small portion of the breast is actually removed, it can be potentially disfiguring. ("New Screening May Replace Breast Biopsy," *McCall's*, May 1991, p. 30)

Note that the two methods are contrasted on the same points: the relative seriousness of the procedure, the time involved, and the amount of trauma to the patient.

In addition to their use in informative prose, comparison and contrast are often used to help a reader decide which item or course of action is best for his or her needs. This type of comparison requires establishing clear CRITERIA, the guidelines or standards against which the items will be compared. For example, you might compare two investments according to their tax benefits, long-term yields, and risk. Chapter 2, "Revising Persuasive Prose," discusses ways to develop criteria.

Classifying

Classifying a group of items helps the reader to see the relationship among different types. As such, classification also involves comparison and contrast. When using classification, check to make sure that the items being compared are sorted into types according to consistent principles. For example, it would be inconsistent to classify investments into tax-exempt, taxable, and high-yield, since two different classification principles have been used simultaneously (tax status and amount of yield).

Excerpt (10), from a discussion of boating safety, classifies personal flotation devices (PFD's) according to their effectiveness for various boating activities and rescue operations.

(10)   Type I: This category is a wearable device, has the greatest required buoyancy and is designed to turn unconscious persons from a face-down position in the water to a vertical or slightly backward position. This PFD is considered by the Coast Guard to be the most effective for all boating activities — especially for offshore cruising, where delayed rescue is a probability.

Type II: This PFD is similar to Type I, but the turning action is not as pronounced and will not turn as many persons under the same conditions as Type I. It's recommended for inshore or coastal cruising. . . .

Type IV: These are throwable PFDs — life rings or flotation cushions are the most common types — that are not designed to be worn. They are for man-overboard situations. ("Personal Flotation Devices," *Motor Boating and Sailing*, July 1991, p. 97)

Using outside sources

Outside sources may include traditional materials such as articles and books, as well as other types of materials such as information from Web sites, interviews, questionnaires, surveys, pamphlets, and brochures. Chapter 2 ("Revising Persuasive Prose") and Chapter 18 ("Editing References to Other Sources") offer some guidelines on how and when to incorporate supporting details from outside sources.

Organizing informative prose

The basic organizational principle to follow in revising informative texts is to move from more general elements to more specific elements. The rationale behind this principle is that readers find details more comprehensible if they are first presented with a general framework for interpreting them. For example, an informative paragraph should usually begin with a controlling idea and then provide supporting details for it. Likewise, a longer informative text such as a report should provide a general overview in the introduction and then move to detail in the body. Chapter 4, "Revising for Paragraph Unity," suggests ways to develop controlling ideas.

Applications

Analyze each passage below for any unsupported generalizations. For each passage, describe at least two types of development techniques that could be used to add supporting detail. Where possible, draft a revision of the text, adding realistic details as needed.

> EXAMPLE — FROM A REPORT ON IMPROVING CUSTOMER SERVICE AT A DEPARTMENT STORE:
> The problem of poor customer service that Harkness Department Store faces lies in the current structure of its pay schedule. The pay schedule is the same one that Harkness used when it first started in 1946. New and more effective pay schedules have been developed in recent years.
>
> ANALYSIS: Terms needing development include *poor customer service*, *current structure of its pay schedule*, and *new and more effective pay schedules*. The writer could quantify or otherwise describe the customer service problem (e.g., "More than 30 percent of our complaints are complaints about employees rather than about products"). The current pay schedule needs to be described (e.g., "For the first three months, new employees receive $4.00 per hour and a 2 percent commission on everything sold over $100 an hour"). Similar details are needed to allow comparison with the other pay schedules, including evidence of their effectiveness.

1.   FROM A REPORT ON INCREASING THE MARKET SHARE OF A MAIL-ORDER COMPANY:
Increased competition in the mail-order industry has left the market saturated. This has left our company with the difficult task of attempting to find new customers in heavily penetrated markets. One problem we face as a result of this is formulating future growth opportunities in the face of such extensive competition.

2.   FROM A REVIEW OF A BOOK ON WOMEN'S PERSPECTIVES ON EXPERIENTIAL EDUCATION:
Section I is entitled "Caring Voices" and contains four chapters centered on feminist ethics, ecofeminism, the midwife teacher, and spiritual empowerment. It's a section that is full of hope, introducing or advancing ideas about how caring relationships with ourselves, others, and the environment can make an impact on experiential education practices.

3.   FROM A REPORT TO THE MANAGER OF A BOOKSTORE: The University Bookstore has created many policies and procedures to keep the textbook operation running efficiently and smoothly. However, many customers do not take these policies seriously, and the store employees do not always stick to them as strictly as they should. As a result, problems have arisen that inconvenience students, faculty, and bookstore staff.

4.   FROM A RÉSUMÉ ENTRY:

Work Experience

Self-employed, working on family farm to help finance college education. Experienced with large amount of individual responsibility. Have experience with management and manual labor.

5.   FROM A RÉSUMÉ ENTRY:

Education

B.S., Accounting, University of Tennessee, expected June 1993. GPA 3.4 on 4.0 scale. Extensive background in all business courses.

6.   FROM AN ARTICLE ON METEOROLOGY IN A POPULAR MAGAZINE: Weather is defined as the atmospheric conditions in a place at a particular time. Climate is defined as the weather in a given place over a long period of time.

7.   FROM A PAMPHLET ON HOW TO PREVENT HEARING LOSS: Sound levels are measured in *decibels*. The sound level of ordinary conversation is 60 decibels.

8.   FROM A USER'S MANUAL FOR WORD-PROCESSING SOFTWARE: The macro function in your program enables you to access a chain of commands by entering one abbreviated command.

9.   FROM A SUPERVISOR'S SEMIANNUAL ASSESSMENT OF AN EMPLOYEE: Ms. Connors has completed a number of important projects with minimal supervision. Her work is usually thorough and accurate. On the other hand, she has missed several staff meetings and often arrives to work late. Some of her accounts have complained of rude behavior.

10.   FROM A REVIEW OF A BOOK ABOUT MANAGING CLASSROOM BEHAVIOR: Chapter 1 discusses several ways to identify behavior problems. It is helpful to the reader because it promotes thought about what should or should not be allowed in the classroom, and why. Tips are also given about which behavior problems to ignore, a difficult decision often faced by beginning teachers.

The book then talks about how to analyze behavior problems. This section has the reader reflect on what misbehavior can and cannot be controlled by the teacher. Identifying behaviors that can be changed is covered next, and instructions for measuring the problem, identifying its antecedents and consequences, and recording the results of changes are given to the reader. These are all useful skills that a teacher will need.

Changing behavior is the next topic covered. Readers are reminded to look at obvious strategies and to remain positive in their approach. The five operations of a behavioral approach are also discussed. This chapter is a natural follow-up to the previous chapter's discussion of analyzing behavior problems since changing the behavior is the reason the behavior is analyzed to begin with.

11.   Identify areas needing development in Appendix 3 ("Brian Carter's Job Application Letter") and revise it accordingly.

12.   Identify areas needing development in Appendix 4 ("Brian Carter's Résumé") and revise it accordingly.

13.   Identify areas needing development in Appendix 5 ("A Definition of Aphasia") and revise it accordingly.

14.   Identify areas needing development in Appendix 6 ("Memo Assessing an Employee's Writing"), Part A, and revise it accordingly.

15.   Identify areas needing development in Appendix 8 ("Research Proposal on 'Erosion at Sawyer Road Park'") and revise it accordingly.

16.   Identify areas needing development in Appendix 10 ("Report on 'Improving Employee Training at Becker Foods'") and revise it accordingly.

17.   Identify areas needing development in Appendix 11 ("Report on 'Security Methods for Ashley's Clothing Store'") and revise it accordingly.

Chapter 2. Revising Persuasive Prose

Persuasive prose takes many forms in professional writing. You may be writing a letter to persuade a potential employer to interview you. You may be writing a proposal to elicit funding from a granting agency or a contract with a new client. Or you may be writing a report to persuade management that an alternative procedure or product is superior to one currently being used. All of these situations require you to convince readers to modify their beliefs or actions. This section reviews some basic elements of persuasion and suggests revision strategies for this type of prose.

Providing a claim, evidence, and interpretation

One approach to persuasion, developed by the philosopher of science Stephen Toulmin, suggests that arguments have three essential elements:

CLAIM: The point that the writer wants to make. Usually this point is a belief or recommendation that the writer wants the reader to adopt, but that is subject to debate.

EVIDENCE: Facts presented in support of the claim. These facts may be drawn from a variety of sources: for example, direct observation, experimentation, or research into secondary sources. Facts are statements that the reader can be expected to believe, as long as they are drawn from reliable sources.

INTERPRETATION: Assumptions that provide a bridge between the claim and the facts, often stated as generalizations or rules. This bridge material provides reasons (sometimes called WARRANTS) that the reader should accept the facts as evidence for the claim.

As an example of these elements, review the argument in (1).

(1) For aircraft, methanol presents some obvious shortcomings. The alcohol content attacks rubber fuel-system parts, meaning that airplanes destined to use methanol would require substantial fuel system changes from the tanks all the way to the engines. They would need larger tanks, too, because methanol contains just half the latent energy of avgas. Fuel consumption would rise drastically, though tests indicated it would not be by 50 percent. What's more, alcohol has the nasty tendency to suspend impurities in the fuel, which would call for extreme care in the handling and filtration of the fuel. (Marc E. Cook, "Fuel for Thought," *AOPA Pilot*, July 1990, p. 68)

This argument can be analyzed as follows:

CLAIM: Methanol is not the best choice for aviation use.

EVIDENCE: Alcohol attacks rubber fuel-system parts; methanol contains half the latent energy of avgas; alcohol suspends impurities in the fuel.

INTERPRETATION: Substantial fuel system changes would be needed; fuel consumption would rise; larger tanks would be needed; fuel would have to be handled and filtered more carefully. (The overriding assumption is that all of these are negative consequences.)

Note that the writer does not explicitly state the overriding assumption; he can safely assume that his readers would view the projected changes as negative.

A persuasive text may need revision because of a problem in any of the three elements:

- The writer may fail to state an explicit claim.
- The writer may fail to present evidence that the reader will believe and that will support the writer's claim.
- The writer may fail to interpret the evidence or may introduce assumptions that the reader disagrees with.

Any of these problems, especially the second two, may cause the reader to question or reject an argument. Chapter 4, "Revising for Paragraph Unity," suggests ways to clarify the claim or main point. The sections below focus on ways to present and interpret evidence effectively.

Developing effective evidence

As stated above, evidence refers to specific facts that substantiate general claims. In the revision stage, the evidence for an argument should be reviewed according to the guidelines below.

Support claims that can be quantified in more than one way. Professional writers are often called on to help the reader choose from among several alternatives. For example, you may find yourself recommending one product or procedure as "more accurate," "more efficient," and "less expensive" than another. All of these terms, however, need further definition to ensure that both you and your reader share the same frame of reference. For example, consider passage (2a), from a feasibility study.

(2a) In the long run, the money spent on a standardized computer system would be saved by the increase in output and decreased teaching/learning time. People who are comfortable with and knowledgeable about computers will more likely be creative and work faster than those who feel restricted by limited computer knowledge.

This passage contains a number of claims that can be quantified in more than one way. As a result, the reader might end up with an entirely different set of assumptions than the writer intended. For example, how long is the "long run"? How much money could be saved?

What does the writer mean by an "increase in output" and "decreased teaching/learning time"? What exactly can the reader expect about greater creativity and faster work? How does one distinguish a "knowledgeable" worker from one with "limited computer knowledge"?

To avoid potential misinterpretation, the writer needs to provide evidence to support the generalizations. Some of this evidence might be included in the paragraph itself, as in the revision shown in (2b).

(2b) In the long run, the money spent on a standardized computer system would be saved by the increase in output and decreased teaching/learning time. We currently spend about $6,000 per year on training costs attributable to the differences among our computer systems. This training cost would be eliminated by standardizing our system. Since the estimated cost of standardization is $13,000, the new system would pay for itself in reduced training costs in just over two years.

Alternatively, the original paragraph might be appropriate in the introduction or conclusion of a report, with evidence supplied in the body of the report.

Support claims based on evidence to which the reader does not have access. As a professional writer, you are responsible for gathering data, analyzing that data, and helping the reader to interpret it for his or her own purposes. While the reader is ultimately most concerned with your claim or recommendation, it's important to let the reader know what evidence you used to arrive at the claim. Does the evidence consist of personal observation? A case study you heard about at a conference? An experiment you read about in a journal? A survey you conducted during a training class? A review of previously published research?

As an example, let's return to one of the claims we examined earlier:

(2c) People who are comfortable with and knowledgeable about computers will more likely be creative and work faster than those who feel restricted by limited computer knowledge.

We have already seen that this passage contains some terms that need further explanation because they can be quantified in several different ways. In addition, the reader needs to know the authority for these claims. For example, perhaps the writer is basing these claims on an experimental study published in a journal. Details about the study can be added to support the general claim, as in (2d).

(2d) People who are comfortable with and knowledgeable about computers will more likely be creative and work faster than those who feel restricted by limited computer knowledge. For example, Markowitz (1996) compared the performance of accounting clerks who had undergone an 8-hour training session on a spreadsheet program to that of clerks given a 1-hour training session supplemented by written materials. The clerks who had an 8-hour training session made an average of 52% fewer mistakes and were able to prepare a spreadsheet 3 times faster than the other group. In addition, the clerks with more training were more likely to incorporate figures into their written reports.

The mechanics of documenting evidence you use from sources other than personal observation are covered in Chapter 18 ("Editing References to Other Sources").

Similarly, passage (3), from a letter of application, also needs to provide evidence for the writer's general claims:

(3)   I possess many personal qualities that are important to any company. Also, I have a good work ethic, which is proven from the many years I have been employed.

Although what the writer says may be true, the passage leaves a number of questions unanswered: what "personal qualities" is the writer referring to? What does the writer mean by a "good work ethic"? Exactly how many years has the writer been employed? And where? Providing evidence would help to convince the reader to accept the generalizations.

Use appropriate evidence. In order to be convincing, evidence must have certain properties.

Authoritative. The evidence should be produced by someone who is established in the subject-matter field; or, if the evidence is produced by a less experienced researcher, it should follow established procedures and lines of research within the field.

Timely. The evidence should reflect the latest available results.

Verifiable. The evidence should be consistent with what another observer would report under the same circumstances. In scientific fields, an experimental study is said to be *replicable* if the researcher provides enough information about methods, materials, and subjects to allow another researcher to replicate, or repeat, the study. In other fields, you may be relying more on secondary sources (research done by others) than on primary sources (research designed by you). You can still ensure that secondary sources are verifiable by using sources that are published or otherwise accessible to the reader.

Representative. The evidence should reflect trends in the research you have examined, not exceptional cases. For example, if only one study out of six suggests that increased training time results in increased accuracy, then the exceptional study should not be presented as the general rule.

Complete. The evidence should be presented thoroughly, with both pros and cons noted. For example, if the Markowitz study noted above found that accuracy rates tended to equalize after a period of one year, this fact should be noted.

Relevant. The evidence should be applicable to the reader's concerns. It is rare to find evidence that is focused on the reader's precise situation. For example, if you are writing a report about whether your particular firm should standardize its computer system, you are unlikely to find published research about your particular firm's current system. However, you may be able to find evidence about firms that are similar to yours in size, mission, employee profile, and so forth.

Interpreting evidence persuasively

Establishing a persuasive link between claim and evidence requires understanding the reader's objectives and beliefs. One way to demonstrate this understanding is to have clearly stated CRITERIA that reflect these objectives. Criteria are the standards against which alternatives are measured. Three criteria that are commonly emphasized by readers in decision-making positions are quality, efficiency, and effectiveness.

To take a concrete example, suppose you work for a small but growing company and are asked to recommend which type of printer the company should purchase. If one of your company's general objectives is to project a more professional image, then this may translate into a specific goal such as "improve the quality and appearance of our proposals and other business communications." In turn, this goal suggests that print quality would be a logical criterion to apply when comparing printers. At the same time, your company's budget constraints may dictate a cost criterion of, say, less than $500. The facts that you collect about various printers will lead to a logical recommendation only when filtered through these criteria. Thus, criteria help both the reader and writer interpret the significance of the evidence and provide a link between the evidence and the claim.

As suggested by this example, it is quite common for two criteria to support conflicting recommendations. For instance, the highest-quality printer may also have the highest initial cost. Understanding the relative importance that your audience attaches to these different criteria can help you to resolve this type of issue. Therefore, you and your potential reader should discuss which criterion should be given priority. Ideally, this discussion should take place before you begin your investigation; however, sometimes the conflict may not become apparent until later in the writing process.

Addressing potential objections

At times, the best choice among alternatives may be obvious to both reader and writer. More commonly, however, your reader may have reservations about your recommendation, questioning either your evidence or your interpretation of it. Instead of ignoring these objections and hoping that they'll go away (which they won't), try to look objectively at your draft, anticipate potential objections to your argument, and respond to them.

For example, suppose your company needs to expand its data processing organization. It has traditionally promoted from within, but you see that this strategy will not work because of the highly specialized nature of computer technology. Therefore, you plan to recommend that the company hire at least some data-processing supervisors from the outside. You have evidence that this strategy will result in a better department and savings in the long run, although the outside hires will command higher initial salaries than internally promoted employees would.

Given the context of this report, it is likely that some readers will resist deviating from a longtime policy of promotion from within. These readers may hold out for an alternative solution which involves, say, retraining current employees so that they can be promoted to supervisory positions in data processing. Advocates of this view may object that outside hires will not understand the corporate culture and that company employees may be demoralized if the promote-from-within policy is ignored. Effectively refuting these objections involves three steps.

Articulate the objection or alternative solution. Define the objection or alternative solution in clear terms so that the reader understands the position being discussed.

Acknowledge the reader's concerns. Recognize the legitimacy of the concerns that would lead the reader to the objection or alternative solution.

Assess the objection or alternative solution according to agreed-upon criteria. Demonstrate that the reader's solution, while meeting some criteria, does not meet other, more important criteria as well as your solution does. This step involves acknowledging any strengths of the reader's view and any shortcomings of your view.

Following these steps can help not only in revising a persuasive text, but also in refining its main recommendation. For example, examining the opposing view in the situation described above might lead the writer to seek clarification on the relative importance of employee morale as a criterion.

Organizing persuasive elements

In general, persuasive texts can be organized either directly or indirectly, depending on the reader's anticipated response. For example, looking back at the paragraph on methanol, note that the writer begins with his claim, then provides evidence and interpretation to support the claim. This direct sequence is effective if you expect the reader to be receptive to (or at least neutral about) your argument. On the other hand, if you expect the reader to resist your argument, you may want to use an indirect sequence, leading into your claim after presenting evidence and interpretation.

Similar principles apply to longer units such as persuasive memos and reports. When writing to a receptive or neutral reader, use a direct structure in which you state your claim or recommendation in the introduction, then move on to provide evidence and interpretation in the body of the text. If you expect your reader to oppose your recommendation, consider an indirect structure that begins with the reasoning to support your claim. In either case, the conclusion of a longer persuasive text should focus on the "big picture," the writer's recommendation.

Applications

1.   Each of the following passages consists of a claim and evidence. For each passage, supply the interpretive material that has been omitted — i.e., the unstated assumption that links the claim and the evidence.

> EXAMPLE: Americans do not understand the causes of the current Palestinian crisis because they do not understand Middle Eastern history.
> ANALYSIS: Understanding the current Palestinian crisis requires an understanding of Middle Eastern history.

a.   Since there are about a million reported cases of child abuse each year, there may be as many new cases of juvenile delinquency each year.

b.   Nothing infuriates the public more than having the price of a necessity increase, so when postal rates go up, most of the public gets upset.

c.   I do not recommend a multi-employer pension plan since such plans protect the rights of employees at the expense of employers.

d.   A color printer comes with this computer; therefore, you should purchase this computer.

e.   Due to the shallow depth of well 52, a thin-walled rod pump is recommended.

f.   There has been a dramatic increase in responsibilities, specifically multiple roles, among women. Such responsibilities can be physically and emotionally trying on women, as well as on their families. Because of these multiple roles, women have an increased chance of becoming depressed.

g.   This device reduces system cost because heat sinks and fans are unnecessary.

2.   Analyze each text below for any unsupported claims. Describe the type of evidence and interpretation that would be appropriate to include in a revised version. Where possible, draft a revision of the text, adding evidence and interpretation as needed.

a.   FROM A REPORT ON GANG PREVENTION: Approaches are needed that provide preventative measures for high-risk individuals and intervention strategies that help both beginning and hard-core members escape destructive lifestyles. Community organization, social intervention, opportunity provision, suppression, and organizational change/development are primary response strategies to gang problems.

b.   FROM A JOB LETTER FOR AN ENTRY-LEVEL POSITION IN MARKETING: The education that I will complete in May includes a variety of business courses related to marketing and communication skills. These courses presented real case studies of problems in major organizations. A strategic study of business aspects was required to find a solution.

c.   FROM A PROPOSAL: I intend to design a well-diversified investment portfolio that will provide the investor with a good return and at the same time offer very little risk. This portfolio will consist of a variety of securities and other financial instruments with varying maturities and risk factors combined to produce a steady stream of income for an indefinite period.

d.   FROM AN ARTICLE: Although some consider homosexuality a disorder, the American Psychological Association does not classify it as such. Homosexuality is not regularly associated with the problems found in psychological disorders.

e.   FROM A PROPOSAL: DataPro has gone through a period of substantial growth over the past years, especially in the area of personnel. However, the company structure is not representative of this. Due to much bureaucracy, employee satisfaction appears lower in the entry-level positions. The turnover rate is increasing, along with negative attitudes of the remaining employees. An analysis of why this is occurring, including recommendations, will improve the company culture and the existing work ethic.

3.   Analyze Appendix 3 ("Brian Carter's Job Application Letter") for any unsupported claims. Describe the type of evidence and interpretation that would be appropriate to include in a revised version. Where possible, draft a revision of the text, adding evidence and interpretation as needed.

4.   Analyze Appendix 6 ("Memo Assessing an Employee's Writing,"), Part A, for any unsupported claims. Describe the type of evidence and interpretation that would be appropriate to include in a revised version. Where possible, draft a revision of the text, adding evidence and interpretation as needed.

5.   Analyze Appendix 8 ("Research Proposal on 'Erosion at Sawyer Road Park'") for any unsupported claims. Describe the type of evidence and interpretation that would be appropriate to include in a revised version. Where possible, draft a revision of the text, adding evidence and interpretation as needed.

6.   Analyze Appendix 10 ("Report on 'Improving Employee Training at Becker Foods'") for any unsupported claims. Describe the type of evidence and interpretation that would be appropriate to include in a revised version. Where possible, draft a revision of the text, adding evidence and interpretation as needed.

7.   Analyze Appendix 11 ("Report on 'Security Methods for Ashley's Clothing Store'") for any unsupported claims. Describe the type of evidence and interpretation that would be appropriate to include in a revised version. Where possible, draft a revision of the text, adding evidence and interpretation as needed.

Chapter 3. Revising Graphics

A GRAPHIC or VISUAL is a numerical or pictorial representation set apart from the prose text. When revising to use graphics effectively, you need to make several types of decisions:

- •Will the reader need a graphic?
- •If so, what type of graphic is best suited for this particular message and audience?
- •How can you design the graphic and integrate it into the surrounding text for the greatest effectiveness?

Deciding whether to use a graphic

Just as there are a variety of graphics types, so there are a variety of reasons for using graphics. As you review your draft, keep the following questions in mind:

Will the reader have to process complex data or large amounts of data?   If so, graphics will probably clarify your presentation.

Do you want the reader to be able to visualize the shape of the data? For example, do you want to show trends over a period of time, or an inverse relation between, say, quality and price among three alternatives? If so, graphics will probably help you to achieve this goal.

Will the reader be intimidated or bored by a number of "gray pages"? For example, are you trying to explain technical material to a nonspecialist reader? Or is the document lengthy? In either case, graphics may help to make the text less threatening and more interesting.

Will the reader be trying to learn from, or use, your text while reading it? For example, are you preparing a set of operating instructions for a piece of machinery? If so, graphics may be easier for your reader to follow than a prose explanation of the process.

The first of these guidelines, the complexity of the data, is probably the most important. Generally speaking, the more individual pieces of data you are presenting, the greater the need for a graphic. At one extreme, the text in (1) is hardly complex enough to need a graphic to supplement or replace it:

(1)   Fewer than 75% of in-home daycare operations are licensed by the state, but 98% of out-of-home daycare operations have state licenses.

The text discusses only one feature (state licensing) of two entities (in-home and out-of-home facilities); therefore, a graphic like Figure 1 would be superfluous.

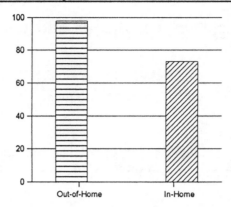

Figure 1. Percentage of daycare operations holding state licenses.

On the other hand, the following passage attempts to discuss two features (the two most common causes of forest fires) of five entities (regions of the United States):

(2)   On the West Coast, the most common causes of forest fire are lightning (31%) and smoking (20%). In the Rocky Mountain states, the most common causes are lightning (64%) and smoking (10%). In the Midwest, the most common causes are debris burning (28%) and smoking (21%). In the Southeast, the most common causes are incendiary burning (39%) and debris burning (19%). And in the Northeast, the most common causes are smoking (26%) and debris burning (21%).

The prose format of this passage makes it hard to see relationships within the information. The patterns in this data are much more apparent in a graphic like Figure 2, where the reader can see all five regions simultaneously. (Note also that the repetitious term "most common causes" needs to be presented only once, in the graphic's title.)

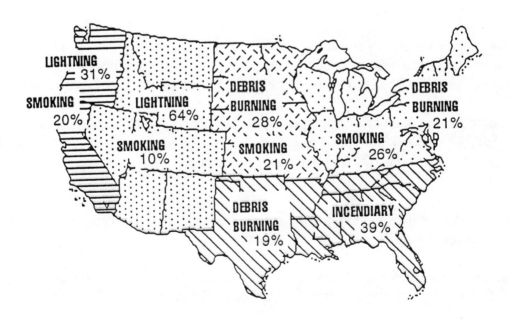

Figure 2. Two most common causes of forest fires by region.

Choosing the best graphic for your audience and purpose

Visuals can be divided into two major categories. ILLUSTRATIONS act as direct representations of the objects they depict. Common types of illustrations are drawings and photographs. INFORMATIVE GRAPHICS, on the other hand, do not directly represent objects. Instead, they convey information about one or more general features of one or more entities. Common types of informative graphics are tables, graphs, charts, and diagrams. (In contrast to a line drawing, which represents the exterior of an object, a diagram explains its operation or composition.)

To decide which type of graphic to use, consider your reader's needs — that is, how the reader will need to use the information displayed in the visual. The list below suggests the types of graphics that are best suited for various reader needs. The graphics are also arranged from least to most complex, depending on how much information each is capable of conveying.

READER'S NEED	GRAPHIC
• to see the surface detail or texture of an object	photograph
• to see the outline and essential parts of an object	line drawing
• to see percentages	pie chart
• to see lines of authority and responsibility	organizational chart
• to see steps in a process	flow chart
• to see values at discrete points in time	bar graph
• to see a continuous trend over time or some other independent variable	line graph
• to see the internal structure of an object	cross-section
• to see how the parts of an object fit together	exploded diagram
• to locate precise values	table

Your audience's level of expertise may also affect your choice of visuals. In general, less technical audiences will find comparatively simple graphics less intimidating and easier to interpret. For example, a nonspecialist reader may be more comfortable with a pie chart or bar graph than, say, a multiple-line graph.

Designing and integrating your graphic clearly

After you have blocked out a rough version of your graphic, start refining it for accuracy, consistency, and clarity.

Use accurate and consistent proportions. When preparing a graph or chart, be sure that the visual proportions in your graph accurately reflect the numerical proportions you are displaying. For example, each increment on the VERTICAL (Y or ORDINATE) AXIS should reflect an equal division of the DEPENDENT VARIABLE. Likewise, each increment on the HORIZONTAL (X or ABSCISSA) AXIS should reflect an equal division of the INDEPENDENT VARIABLE.

Scale is also important; according to the American Psychological Association, a realistic scale can be achieved by using a ratio of 2:3 for the relative lengths of the ordinate and abscissa. For example, if the vertical axis is 3 inches long, the horizontal axis should be 4.5 inches.

Label all parts clearly. If you are preparing a line or bar graph, check that you have identified the units for the x and y axes. Multiple-bar, stacked-bar, or multiple-line graphs should also include a LEGEND that identifies the different items being graphed. On a pie chart, label the percentage value of each wedge with a horizontal label; describe each wedge with either a horizontal label or a legend.

Give each graphic a label and title and, if appropriate, a caption. Reports and articles typically use two different numbering and labeling sequences for tables and for figures (all non-tabular visuals). Be sure to label each type of graphic to reflect one of these series (e.g., Table 1, Table 2; Figure 1, Figure 2). Each graphic should also receive a TITLE, a phrase that describes the graphic as a whole. You may also want to include a CAPTION, one or two complete sentences that help the reader interpret the graphic and see its relevance. Put another way, the title describes the topic of the visual; the caption states the visual's main idea. A caption is especially useful when a graphic is separated from its discussion in the text (for example, when the graphic appears in an appendix).

Integrate each graphic into the text. Integrating a graphic into the text involves placing it effectively and referring to it clearly. As a rule, you will want to place a visual as close as possible *after* its first mention in the text, ideally on the same page or the following one. In the text leading up to the visual, introduce the visual by mentioning its label and referring to it in an informative way. Rather than simply telling the reader to "See Figure 3," help the reader to interpret the visual, as in examples (3a)-(3c).

(3a) Nearly 35% of all cases are curable, shown in Figure 2.
(3b) Only one subject identified all of the stimuli correctly (see figure 4).
(3c) Table 1 summarizes our sales figures for the past two years.

You should also follow up the graphic with an interpretive comment in the text; for example, *As shown in Table 1, our sales typically drop during the first quarter.*

Applications

EXAMPLE: Evaluate the following table's effectiveness for displaying the information it contains. Revise as necessary.

Table 1. Daily Percentage of Weekly Vehicle Use on Cades Cove Road.

Day	Mon.	Tues.	Wed.	Thur.	Fri.	Sat.	Sun.
%	6	15	6	6	12	29	26

ANALYSIS: The table is adequate if the writer's goal is merely to state the exact percentage associated with each day, but otherwise it is ineffective in helping the reader to visualize any trend in the shape of the data. To compare day-to-day changes, the writer could present the data in a bar or line graph, as shown here.

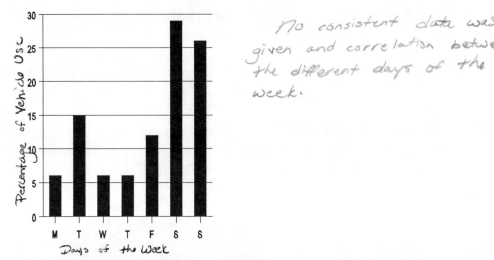

No consistent data was given and correlation between the different days of the week.

1.   Further revise the bar graph above so that it has a more realistic scale. Also supply a label for the y axis and a title and caption for the graph.

2.   Try putting the information in the Example problem into a pie chart. Why is a pie chart problematic for this data?

You'd have to make a pie chart for each day of the week. Pie charts are generally used for diffe

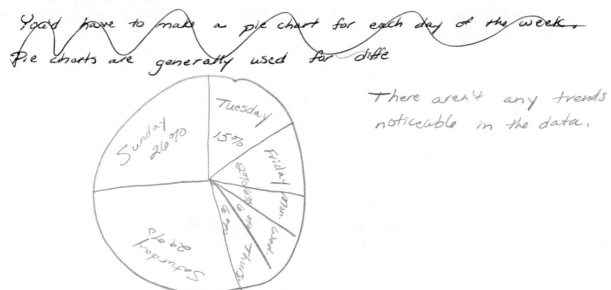

There aren't any trends noticeable in the data.

3.   Provide a caption for the forest fire graphic on page 24.

4.   Revise the pie chart below entitled "Causes of Forest Fires, Rocky Mountain Region." In your revision, consider a more effective way to arrange the segments and to display the legend. Explain and justify your changes.

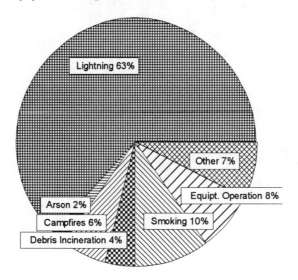

5. Examine the graphic below and revise it after answering the following questions.

a. Based on a brief visual examination of the line, between which years did the greatest decline in reported cases occur? The greatest increase? *85-87 greatest decline*
81-83 greatest increase

b. Now examine the values on the y axis. Do your answers remain the same? *no*

c. Explain the problem with the y axis, and revise the graph to correct this problem. *The y axis is inconsistent on the way the numbers rise*

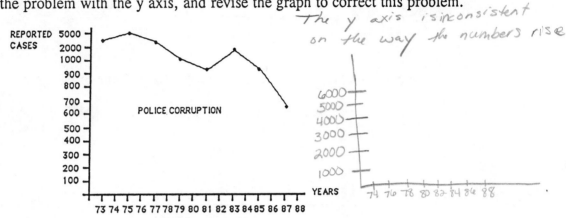

6. Information in graphs and tables can be translated verbally as "if unit X, then property Y." In a graph, the independent variable is usually represented along the horizontal axis from left to right. The dependent variable is represented along the vertical axis.

a. How does the following graph, "Toxin Level in Eggs," fail to conform to this general rule?

b. Re-sketch this graph so that it conforms to the general rule.

c. Revise the following caption so that it both summarizes and interprets: "This graph shows the ratio of toxin in a bird's diet to the toxin detected in the bird's eggs."

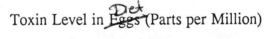

Toxin Level in ~~Eggs~~ *Diet* (Parts per Million)

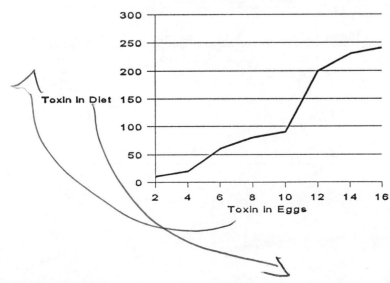

7. Examine the graphics below, "Percentage of Total Catch, Lake Nacogdoches." The information being presented is as follows:

1986: 16" fish = 14% of catch; 12" fish = 28% of catch
1987: 16" fish = 18% of catch; 12" fish = 42% of catch

a. Based on the general principle described in Application 6, what is the problem with the way this information is presented in the two graphs below?

It gives inaccurate information about fish of different lengths

b. Revise the graphics, combining them into one line or bar graph.

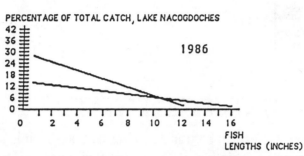

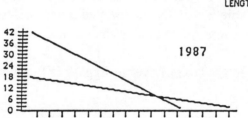

PERCENTAGE OF TOTAL CATCH, LAKE NACOGDOCHES

1986

FISH LENGTHS (INCHES)

1987

8. The following table appeared in a report emphasizing the importance of job satisfaction. Write an introductory paragraph and a follow-up paragraph to help the reader interpret the table.

Table 1. Sources of Life Satisfaction

	Total	Supervisory	Middle Mgmt.	Executive
Home Life	40.1%	36.8%	46.1%	37.8%
Outside Interests	9.8	17.4	11.3	7.9
Career	50.1	45.8	42.6	54.3

A study was done to see how people rate their career, home life, and outside interests above one another.

9.   The following text comes from the summary section of a report recommending whether a company should expand its marketing efforts into Japan, Canada, or Mexico. Develop a graphic (including a title) that could substitute for or supplement this passage. Include introductory and follow-up text to integrate the graphic into the surrounding report.

Japan has traditionally had high tariffs; however, they are lower today. Non-tariff barriers are in the form of controlled distribution systems and cultural differences. U.S. and Japan have a favorable exchange rate. Japan's transportation systems are extensive and efficient. U.S.-Japan trade agreements exist.

Canada has low tariffs except for Ontario. Cultural barriers are present due to the dual cultural heritage. Canada's transportation systems are well developed but limited because of a low population density. The U.S.-Canadian exchange rate is favorable. Trade agreements are present.

Mexico has high restrictive import policies. Cultural barriers are present. The U.S. and Mexico exchange rate is unfavorable because the peso is undervalued, causing imports to decrease. Mexico's transportation systems are not fully developed. Trade agreements are present but restricted to certain products.

10. Analyze the effectiveness of the following graphic. If appropriate, revise the graphic. Also, add interpretation to the surrounding text to help the reader understand the main point of the graphic.

Table 1. Reason for choosing a particular running shoe

Reason	Total	Women	Men
Price	33.0%	39.2%	28.1%
Selection	23.5%	15.7%	29.7%
Quality	11.3%	9.8%	12.5%
Convenience	6.1%	2.0%	9.4%
Satisfaction	13.0%	11.8%	14.1%
Other/Do not know	13.1%	21.5%	6.2%

11. The following passage is from a report on the accounting procedures for vehicle usage at a company. Design a graphic that could supplement or replace the text.

CURRENT PROCEDURES

This section provides a step-by-step outline of the current procedures.

Step 1. Each time a vehicle is used, the user records the mileage and appropriate account number on an individual vehicle usage report that is kept in the vehicle. The account number referred to is the account number of the project that the vehicle use is to be charged to. All expenses such as gas, oil, and other maintenance are also recorded on this report.

Step 2. At the end of each month, the individual vehicle usage report for each vehicle is sent to the area office that the vehicle is assigned to. An Office Supervisor or a Customer Account Representative transfers the information from the individual vehicle usage reports to a monthly vehicle usage report. This report has a line to summarize the usage of each vehicle assigned to that area office. A copy of this report is then sent to the main accounting office.

Step 3. Also, in the area office, the information on the monthly vehicle usage report is entered onto the vehicle usage screen of the intracompany computer system.

Step 4. The individual vehicle usage reports are then sent to the division's general accountant at the division office. The general accountant verifies all information and arithmetic on the reports, and then transfers the information to a spreadsheet. Each row of this spreadsheet includes the month's mileage, gas expense, oil expense, and other maintenance expense for a vehicle in the division.

Step 5. A copy of this spreadsheet is given to the Division Engineer, who uses it to check on vehicle charges to specific projects.

Step 6. A copy of this spreadsheet is also given to the Vehicle Maintenance Coordinator, who uses it to keep track of vehicle maintenance expenses.

Step 7. A person in accounting at the main office compares the data on the monthly vehicle usage report with the data on the vehicle usage screen for each division. This data will then flow through to be charged to appropriate accounts.

12.   The following table shows a company's annual sales and the percentage of those sales attributed to a Future Sales program (a program allowing retailers to receive a 7% discount on orders placed six months in advance).

    a.   Construct a graph that will help the reader visualize the trend in this data.

    b.   What is the relationship between the direction of annual sales and that of Future Sales? Construct a title and caption to convey this relationship.

	Annual Sales	Percent in Future Sales
1991	$500 million	60%
1992	$700 million	60%
1993	$870 million	60%
1994	$600 million	40%
1995	$800 million	60%
1996	$1 billion	60%
1997	$800 million	45%
1998	$2 billion	85%

13.   Identify at least one place in Appendix 5 ("A Definition of Aphasia") that would be clearer if a graphic were added, and sketch or describe the type of graphic that would be appropriate.

14.   Identify at least one place in Appendix 8 ("Research Proposal on 'Erosion at Sawyer Road Park'") that would be clearer if a graphic were added, and sketch or describe the type of graphic that would be appropriate.

Chapter 4. Revising for Paragraph Unity

Paragraphs provide readers with important signals about how text material is organized. Readers expect the information in a paragraph to relate to a central idea (often called a CONTROLLING IDEA). They also expect each paragraph to contain an appropriate amount of detail to support its central idea — enough to be convincing, but not too much to be manageable. When a paragraph meets these expectations, the effect is one of PARAGRAPH UNITY. As you revise the paragraphs in your draft, you can unify them by checking for two problems: lack of a controlling idea, and paragraphs that contain too much detail to be presented as a unit.

Start with a controlling idea

Paragraphs that lack unity often include a series of details but no explicit statement of the general point that the details support. This problem exists in excerpt (1a), from a letter of application. The writer was applying for a job as a product representative with a ski company. (Sentences have been numbered for easier reference.)

(1a) LACKING UNITY: [1]My Business Administration degree is highlighted with a concentration in Marketing. [2]After completing my degree, I hope to continue cross-country skiing and at the same time work within the industry. [3]Because I've been in the ski-racing community for more than 10 years, starting in high school with USSA and then moving to college racing in the NCSA, I have met and made friends with many fellow skiers. [4]This may open up new opportunities for marketing Kniessl Skis.

One way to spot the lack of unity in this paragraph is to compare how it begins and how it ends. Sentence (1) prepares the reader for supporting details about the writer's academic preparation for the job (for example, courses that the writer has taken, projects that he has completed in these courses, and so on). However, sentence (2) discusses the writer's plans for after graduation. Sentences (3) and (4) take yet another direction, discussing how the writer's activities in the ski-racing community would prepare him for the job he is applying for. In short, the paragraph begins on one topic and ends on another, reflecting its lack of unity.

To revise this paragraph for greater unity, the writer could begin by determining what topic most of the paragraph deals with, and then construct a controlling idea that conveys this topic. The writer should probably set aside the first sentence for now and try to relate it to a different set of details in a different paragraph. He could then construct a controlling idea for the remaining details, all of which deal with the writer's skiing activities. Example (1b) shows one possibility.

(1b) UNIFIED: My background in cross-country skiing will complement my academic preparation in Marketing. Because I've been in the ski-racing community for more than 10 years, starting in high school with USSA and then moving to college racing in the NCSA, I have met and made friends with many fellow skiers. This may open up new opportunities for marketing Kniessl Skis.

This version begins with a controlling idea that is supported by the details in the paragraph and that appeals to the reader's concerns about the writer's qualifications.

Sometimes the writer's controlling idea will set up certain expectations on the reader's part that are then unmet by what follows. This problem occurs in (2a), from the "Methodology" section of a proposal. (Again, sentences are numbered for easier reference.)

(2a) [1]To reach the objective outlined in this proposal, both primary and secondary research will be used. [2]Secondary research such as journal articles and previous studies will support information gathered through primary research. [3]Case studies and interviews will be the source of primary research. [4]We will be tracking children from their start in a special needs classroom through mainstreaming into a regular classroom. [5]Listed below are the names and job descriptions of those who will be interviewed . . . .

Sentence (1) provides a (potentially) clear controlling idea. After reading it, the reader expects to hear first about primary research and then about secondary research. However, these expectations are violated because the writer discusses secondary research first. It would be relatively simple to change the order of the subjects discussed so that they reflect the order set up in sentence (1), as in (2b).

(2b) [1]To reach the objective outlined in this proposal, both primary and secondary research will be used. [2]Case studies and interviews will be the source of primary research. [3]We will be tracking children from their start in a special needs classroom through mainstreaming into a regular classroom. [4]Listed below are the names and job descriptions of those who will be interviewed . . . . [5]The information gathered through primary research will be supported by secondary research such as journal articles and previous studies.

Group details into manageable units

A second reason that some paragraphs lack unity is that they contain details that support more than one controlling idea. Often such paragraphs are relatively long (e.g., more than a half page of double-spaced text). Frequently, such a long unit needs to be reorganized into two paragraphs, each with its own controlling idea. For example, consider (3a), from a job letter by a student applying for a recruiter/trainer position.

(3a) LACKING UNITY: [1]At the end of November, I will be graduating from State University with a Bachelor of Business Administration degree. [2]While attending State, my concentration and emphasis has been Human Resource Management. [3]I have supplemented my general business coursework, which includes marketing, finance, accounting, economics, and composition, with courses in compensation and benefits, human resource development-training, staffing, and strategic planning. [4]Through these classes, I was able to develop an organization's compensation/benefit plan, strategically plan for an organization's staffing needs, and develop a training program. [5]My time at State was also highlighted with extracurricular activities. [6]As the vice president of our

Society for Human Resource Management, I helped to organize meetings and coordinated the annual Dress for Success show. [7]I have also been active in the Business School's freshman mentorship program, which raises freshman interest in business-related programs and organizations at State.

This student has some strong qualifications and provides convincing details about them. However, the lack of paragraph unity makes it hard to see the forest for the trees: she needs to clarify the main trends that the reader should see among the details. There is simply too much detail for the reader to absorb in one unit.

Reading back through this paragraph, we can see that the writer discusses two aspects of her career at State University. Sentences (1)-(4) all deal with the writer's academic preparation for the job — that is, her formal coursework. Sentences (5)-(7), on the other hand, focus on extracurricular activities. This change in topic suggests the need for two distinct paragraphs.

The next step is to give each paragraph its own controlling idea, one that ties together the details and relates them to the writer's purpose. Since the writer's purpose is to show that she qualifies for an interview, the controlling idea should reinforce this aim, as in (3b).

(3b) UNIFIED: My degree in Business Administration, which I will receive this November from State University, has given me a strong background in Human Resource Management and its applications to recruiting and training needs. As a major in Human Resource Management, I have supplemented my general coursework in marketing, finance, accounting, economics, and composition with courses in compensation and benefits, human resource development-training, staffing, and strategic planning. Through these classes, I was able to develop an organization's compensation/benefit plan, strategically plan for an organization's staffing needs, and develop a training program.

My extracurricular activities have also required skills in organizing and motivating people. As the vice president of our Society for Human Resource Management, I helped to organize meetings and coordinated the annual Dress for Success show. I have also been active in the Business School's freshman mentorship program, which raises freshman interest in business-related programs and organizations at State.

Now each paragraph focuses on one topic — the first on the writer's academic qualifications, the second on job-related skills developed through her extracurricular activities. Each paragraph begins with a controlling idea, and the previously overwhelming amount of detail has been divided into two, more manageable, units.

Likewise, paragraph (4a), from a draft of an article on multiple roles as a factor for depression in women, needs to be reorganized into several paragraphs with clearer controlling ideas.

(4a) LACKING UNITY: [1]Bird and Ross (1994) found that relative to paid work, housework provides more autonomy, as measured by the opportunity to decide what to do and how to do it. [2]However, they also found housework to be more routine, to be less intrinsically gratifying, and to offer fewer extrinsic rewards. [3]Bird and Ross (1994) also examined the relationship between the women's work conditions and sense of personal control. [4]They report that routine, ungratifying, and unrewarding work accounts for homemakers' lower sense of personal control relative to employed women. [5]Self-directed work, work that allows for the use of initiative and independent judgment, is psychologically beneficial whether it is found on the job or in the home. [6]Employed wives report greater time pressure and responsibility than homemakers (Lennon, 1994). [7]Lennon (1994) found that housework as a full-time activity is reported to be more autonomous, more subject to interruption, more physically demanding, and more routine than the average paid job, while paid work involves more time pressures and greater responsibility than housework. [8]In assessing the relationship of work conditions to depressive symptoms, Lennon (1994) found that housewives' symptoms significantly exceed those of employed wives. [9]The extent to which the worker is responsible for things outside her control and the amount of routine her work involves are associated with greater depressive symptoms, regardless of her work status (Lennon, 1994). [10]In comparing the two, homemakers obtain a certain benefit from having less responsibility, and employed wives seem to benefit from less routine work. [11]In balancing out the two, employed wives and homemakers exhibit similar levels of depressive symptoms.

Paragraph (4a) creates a "rambling" effect because the writer does not clarify at the beginning exactly what it is she is comparing, and she does not present a generalization about her subject at the beginning. She needs to bring a controlling idea out earlier so that the reader can use it to interpret what follows. In addition, because its length and amount of detail, (4a) is a good candidate for being divided into at least two paragraphs, perhaps involving some re-ordering of the sentences. Try blocking out a revision that accomplishes these changes.

Applications

1.   For each paragraph below, develop a topic sentence that will give the paragraph greater unity by incorporating a controlling idea.

>   EXAMPLE — FROM A LETTER OF APPLICATION FOR AN ENTRY-LEVEL ACCOUNTING POSITION: While at State University, I have been employed in the Office of Financial Aid as an office assistant. My employment has included duties such as computer work on Lotus, counting and reporting cash, and recording confidential information. In the coming weeks, I am scheduled to work with the general ledgers of the university in order to reconcile past entries with corresponding documentation of the College Work Study Program.
>
>   REVISION: My part-time work during college has already enabled me to apply many of the skills required of an accountant. While at State University, . . .

>   a.   FROM THE METHODOLOGY SECTION OF A PROPOSAL: The research into this project will require me to work with the Dixie Cash Register Company as well as IBM. The two companies work together to provide businesses with the best system for their needs. I will also observe different systems at work in various stores in the area. This must then be incorporated into the best system for Belmead's. I will then analyze present costs and future savings that the system will provide.

>   b.   FROM A LETTER OF APPLICATION TO A CIVIL ENGINEERING FIRM: My work experience has consisted mainly of farmwork. I helped to manage a farm of nearly 1,000 acres employing from four to fifteen people, depending on the season of work. Helping to design and build large barns and houses gave me the insight to be a structural engineer. This past summer I worked for the State University Police Department, where I learned to work with many different people.

c.   FROM AN EXECUTIVE SUMMARY: Marks & Spencer faces a significant threat from the government. Government intervention has been an external issue in the past because of the high tax rate they placed on Marks & Spencer. The government can continue to be a threat in the future. Marks & Spencer has three threatening competitors. Supermarket chains with high food sales are one threat. Retailing firms that imitate Marks & Spencer but sell lower-quality goods are a threat. The final competitive threat is American-style department stores with less expensive, fashion-oriented goods.

d.   FROM THE REVIEW OF THE LITERATURE IN A CHEMISTRY ARTICLE: Limblad and Pakkanen used cluster models to describe the interaction of Hcl with non-polar surfaces of y(gamma)-alumina [1993]. Other researchers have modeled interaction at Lewis acid sites Hirva and Pakkanen, 1992; Fleisher et al., 1992] and Bronstead sites on alumina [Kawakami and Yoshida, 1985].

2.   Suggest ways to divide each passage below into two or more paragraphs, developing a controlling idea for each paragraph.

a.   [1]This book offers a chance for a new generation of parents, teachers, grandparents, and siblings to discover the rewards and the necessity of reading aloud to children. [2]The layout of the book is very well organized; however, it should be, considering this is its fourth edition. [3]The book contains no pictures or graphics that would help keep the reader's attention. [4]The style of writing the author uses fits well with the audience. [5]He mixes factual information with personal stories on the same page to keep the readers from getting bored with the material. [6]The author does an excellent job targeting his broad audience by not using overly technical words that could cloud their understanding. [7]One area that is weak is that none of the chapters have any sense of closure. [8]Not a single chapter contains a conclusion where all the information is summarized for the reader. [9]The last chapter in the book tries to bring all the information together but only covers about half of the chapters throughout the book. [10]Overall, this book is an excellent resource for anyone that interacts with children. [11]I feel that this book should be on every coffee table in America.

b.   [1]Ethyl alcohol is the least toxic member of the alcohol family. [2]Medically, it is classified as a hypnotic or sleep producer. [3]Commercial ethyl alcohol, employed in alcoholic beverages, is used in its natural state of low toxic levels and a maximum of 95% concentration of ethyl alcohol. [4]The amount and quality of commercial ethyl alcohol are carefully regulated to protect the consumer, who consumes the alcohol internally. [5]Industrial ethyl alcohol, such as that used in solvents, is not nearly as regulated, since human consumption is not involved. [6]Industrial alcohol is used at any concentration to 100% ethyl alcohol. [7]However, to ensure that industrial grade alcohol is not used as commercial alcohol, a denaturant agent is included in the industrial alcohol. [8]There are two types of denaturant agent available for use. [9]The first type increases the natural toxic level of the alcohol from mildly poisonous to extremely poisonous. [10]The second type of denaturant causes a very poor taste in the alcohol, such as the alcohol used in perfumes.

c.   [1]The natural home environment of a child prenatally exposed to cocaine can be more harmful than the exposure to drugs (Williams & Howard, 1993). [2]The environment often includes one or more of these problems related to drug abuse by the parent(s): neglect, abandonment, postnatal exposure to drugs, the return to drug abuse, and child abuse. [3]Neglect and abandonment occurs when the mother leaves to search for drugs, returns to prostitution to make money to buy drugs, or uses the grocery money to buy drugs (Williams & Howard, 1993). [4]Postnatal exposure to drugs has been reported "where infants passively inhaled smoke from free based cocaine or ingested cocaine through breast milk" (Williams & Howard, 1993, p. 68). [5]The potential for child abuse in families where drugs are abused is high: "A task force in New York City found that there was a 72 percent increase in child abuse from1985-1988 as a result of drug abuse, primarily crack" (Williams & Howard, 1993, p. 68). [6]Mothers who were abused tended to abuse their own children. [7]Combine this with frustration from a newborn, frustration from poverty, and frustration with the situation, and it is more likely that abuse will occur. [8]When the mother returns to drug abuse, she leaves the child "in a chaotic environment that puts them at risk of developing all sorts of problems: learning difficulties, hyperactivity, even tendencies to violence" (Waller, 1994, p. 30). [9]The parenting ability of cocaine-using mothers may be compromised because the use affects the consistency and predictability of the parental behavior toward the child (Griffith, 1992). [10]The home environment that these children grow up in is chaotic and unstable. [11]This unstable environment and the lack of interaction with parents can lead to language delay because the parent is not available to interact and teach the child language.

d.   [1]Street gangs (a group or association of three or more persons who may have a common identifying sign, symbol, or name, and who individually or collectively engage in, or have engaged in, criminal activity, or as juveniles commit an act that if committed by an adult would be a criminal act) have been a part of America's urban landscape for most of the country's history and subject of research since at least the 1920's. [2]But most street gangs in the first third of the century were small groups involved in delinquent acts or relatively minor crimes, primarily fights with other gangs. [3]Today there are many more different types of street gangs. [4]Individual members, gang cliques, or entire gang organizations traffic in drugs, commit shootings, assaults, robbery, extortion, and other felonies, and terrorize neighborhoods. [5]Until recently, research on gangs centered on exploring reasons for gang formation and participation, with a related emphasis on public policy that deters vulnerable youths from joining gangs. [6]But the destruction and fear generated by today's street gangs have elevated the importance of research on the effective community and criminal justice responses to them (Johnson 1995). [7]Street gang membership encompasses all races from most socioeconomic levels and is not limited to larger cities. [8]Do not think that joining a gang is just a phase. [9]Older gang members use newer gang members or people looking to get into the gang by having them take most of the risk. [10]Although gang presence in large cities can be traced back well over 200 years, in the past decade street gangs have grown in both size and sophistication (Owens 1993).

e.   [1]According to Peter Hodges, in the past 10 years the sheer volume of direct-mail selling has more than doubled. [2]Also, many companies have faced reduced response rates. [3]The reason for the reduced response rates is more likely caused by the increase in the volume of direct-mail selling. [4]With more companies involved in this type of industry, the industry pie must be cut into more pieces; thus consumers are overwhelmed with more choices on which to spend their money. [5]Because most direct-mail companies have been having low response rates, this problem is considered widespread and not necessarily concentrated on just a few companies. [6]The low response rates are due to the saturated catalog market of so-called "junk mail." [7]Therefore, we need to make drastic changes in our strategies in order to increase response rates. [8]Generally, the companies that become more innovative in the direct-mail industry and whose innovations are accepted by consumers will gain the most market share. [9]These innovations or changes in our strategies can come in the form of changes in television advertising, shipping and handling costs, new-customer gift certificates, contests, and a toll-free number.

3.   Revise Appendix 1 ("Memo to Residents") for paragraph unity.

4.   Revise Appendix 8 ("Research Proposal on 'Erosion at Sawyer Road Park'") for paragraph and section unity and use of unifying headings.

5.   Revise Appendix 9 ("Executive Summary on 'ProWear Shoes'" for paragraph unity.

6.   Revise Appendix 10 ("Report on 'Improving Employee Training at Becker Foods'") for paragraph unity.

7.   Revise Appendix 11 ("Report on 'Security Methods for Ashley's Clothing Store'") for paragraph and section unity and use of unifying headings.

8.   Revise Appendix 12 ("A Discussion of Prenatal Exposure to Cocaine") for paragraph unity.

Chapter 5. Revising for Cohesion

A text has COHESION if its parts are unified (i.e., the parts relate to one topic and purpose) and the movement from part to part is easy to follow. One powerful revision tool for creating greater cohesion is the GIVEN-NEW PRINCIPLE, which assumes that a sentence contains two information units. GIVEN INFORMATION (sometimes called OLD INFORMATION or the TOPIC) is information that the writer assumes is already known to the reader. NEW INFORMATION (sometimes called the COMMENT) is information that the writer assumes is not known to the reader. Studies have shown that texts are easier to process and remember if they adhere to the following pattern:

- Given information appears in the SUBJECT slot (i.e., in a noun phrase at the beginning of the sentence);
- New information appears in the PREDICATE slot (i.e., in the second part of the sentence, containing the verb and its object(s) or complement).

For example, passage (1a) violates the given-new pattern by presenting information in a sequence of given-new (in the first sentence), new-given (in the second sentence), new-given (in the third sentence). (Given information is underlined; new information is italicized.)

(1a) DRAFT: The Model 101B *is the result of new processing technology. A number of exciting advances are offered by* this product. *A speedier relay time, improved noise reduction, and a Synctron readout mode* are among its new capabilities.

Now compare (1b), which adheres to a sequence of given-new, given-new, given-new.

(1b) REVISED: The Model 101B *is the result of new processing technology.* This technology *allows us to offer you a number of exciting advances.* The 101B's new capabilities *include a speedier relay time, improved noise reduction, and a Synctron readout mode.*

Three types of given-new patterns

The given-new pattern can be implemented in one of three basic ways: an AB:BC pattern, an AB:AC pattern, or a combined pattern.

The AB:BC pattern. Example (1b) above adheres to an AB:BC pattern, where the new information in the predicate of the first sentence becomes the given information in the subject of the second sentence. This pattern is analyzed in more detail here.

(1b) The Model 101B *is the result of new processing technology.*
         A   B

This technology *allows us to offer you a number of exciting advances.*
     B   C

The AB:BC pattern also links the second and third sentences: the new information in the predicate of the second sentence (*a number of exciting advances*) is paraphrased as given information in the subject of the third sentence (*new capabilities*). Thus passage (1b) follows a progression of AB:BC, CD:DE.

The AB:AC pattern. Another way to implement the given-new pattern is to repeat the given information as the subject of each sentence, adding new information in each predicate slot. This pattern is illustrated in (1c). (Again, given information is underlined, and new information is italicized.)

(1c) The Model 101B *is the result of new processing technology.* This product *offers a number of exciting advances.* It *has a speedier relay time, improved noise reduction, and a Synctron readout mode.*

The combined pattern. The AB:BC and AB:AC patterns may be used in the same paragraph, as illustrated in (1d). Here, the first two sentences are linked by the AB:AC pattern. The second and third sentences are linked by the AB:BC pattern.

(1d) The Model 101B *is the result of new processing technology.* The 101B *offers a speedier relay time, improved noise reduction, and a Synctron readout mode.* These features *allow you to compile test results quickly and accurately.*

Linguistic strategies for creating a given-new pattern

Four revision strategies are especially useful for creating a given-new pattern.

Repeat a noun phrase to signal given information. One way to reinforce given information is to repeat a noun phrase. For example, in (1b) (repeated below), *technology* is new information in the first sentence and given information in the second sentence.

(1b) The Model 101B *is the result of new processing technology.*
　　　　　A　　　　　　　　　　　　　　　　　B

This technology *allows us to offer you a number of exciting advances.*
　　　　B　　　　　　　　　　　　　　　　　C

Use a synonym to signal given information. Another way to reinforce given information is to use a synonymous noun phrase. This strategy is used in (1c), repeated here.

(1c) The Model 101B *is the result of new processing technology.*
　　　　　A　　　　　　　　　　B

This product *offers a number of exciting advances.*
　　　A　　　　　　　　　C

The phrase *this product* is synonymous with *the Model 101B*, and this relationship allows the reader to link the given information in the two sentences.

Use a pronoun to signal given information. Since a pronoun is, literally, a word that stands "for a noun," you can use a pronoun to refer to an earlier noun phrase. Two types of pronouns can signal given information. PERSONAL PRONOUNS occur in the same positions that noun phrases do (e.g., in subject position). In (1c) (repeated below), a personal pronoun indicates given information in the second sentence:

(1c) <u>The Model 101B</u> *is the result of new processing technology.* <u>It</u> *has a speedier relay time, improved noise reduction, and a Synctron readout mode.*

DEMONSTRATIVE PRONOUNS (*this, that, these*, and *those*) are used to modify another noun; they appear at the beginning of a noun phrase (i.e., before the noun and any adjectives). Demonstrative pronouns indicate given information. For example, in (1b) (repeated below), the demonstrative pronoun marks the subject noun phrase *this technology* as given information.

(1b) <u>The Model 101B</u> *is the result of new processing technology.* <u>This technology</u> *allows us to offer you a number of exciting advances.*

Vary active and passive voice to focus given and new information. Varying active and passive voice changes the order of the subject and predicate noun phrases. (For a review of how to identify the passive voice, see Chapter 10.) By being aware of these changes, you can revise for greater cohesion. For example, (1e) violates the given-new pattern because the old information (*this mode*) follows the new information.

(1e) <u>The model 101B</u> *features a Synctron readout mode.*
      A   B

     Test results can be compiled much more accurately using <u>this mode</u>.
              C   B

Changing the second sentence to active voice, as in (1f), re-positions *this mode* into the subject slot, a more appropriate place for given information.

(1f) <u>The model 101B</u> *features a Synctron readout mode.*
      A   B

    <u>This mode</u> *allows the operator to compile test results much more accurately.*
     B   C

Applications

Revise each passage below for greater cohesion, using the AB:BC pattern, the AB:AC pattern, or a combined pattern.

EXAMPLE: The BELOROBIC process produces about 6 cubic feet of gas for every pound of DOC removed. The participating industry qualifies for government fuel credits by using this gas as an industrial motor fuel. These credits result in a substantial savings over a period of just one year.

REVISION: For every pound of DOC removed, the BELOROBIC process produces about 6 cubic feet of gas. By using this gas as an industrial motor fuel, the participating industry qualifies for government fuel credits. These credits result in a substantial savings over a period of just one year. (AB:BC)

1. Some fruits contain more oil than others. The amount of oil in a fruit determines the fruit's caloric value. Birds' fruit preferences correlate with this caloric value. In other words, one way that birds choose the foods they eat is through caloric value.

2. Many manufacturers provide information and recommendations on filter performance. Producers will be able to select the appropriate system for their orchards by using this information carefully.

3. I have proposed that a standardized crew cycle be developed and used in the production of future mission timelines. The amount of time and labor currently required to produce a mission timeline would be greatly reduced by the use of such a crew cycle.

4. A five-year-old thoroughbred mare was referred for (1) a sesamoid fracture of the right rear leg and (2) and upper respiratory problem characterized by a cough and water running out of the mare's nostrils when drinking. The client had first indicated both these problems four months earlier.

5.   Our proposal outlines the basic plan for this project. We will submit more specific plans after discussions with the client.

6.   According to data collected from surveys, many homeschoolers feel that they have a closer relationship with their family than they would if they went to school everyday. High self-esteem and productivity may result from this closeness.

7.   Cross-contamination is the term used to denote the passing of germs and bacteria from one place to another. It is believed that cross-contamination is the major means by which bacteria are transported in commercial food service, particularly in restaurants where the pressure is great to produce quality products in minimal time. For example, it seems to be a waste of time to stop and wash hands or get a clean spoon for every tasting during a rush period. However, product contamination is frequently the result of such activities.

8.   Effective instruction in foreign accent reduction involves effective communication between clinician and client. Only when the clinician and client can interact comfortably will effective communication be achieved. Comfortable interaction occurs when the clinician is sensitive to cultural differences in communication.

9.   The manufacturer's representatives must be thoroughly knowledgeable about the products they sell. In addition, both their customers and the manufacturers they represent must trust them.

10. The use of composite materials in an aircraft reduces the amount of drag on the airplane. The number of seams on the surface of the aircraft is reduced by the use of composites because composites are produced in large panels. The need for rivets is minimized by the use of composites because most composite structures are joined by adhesives.

Chapter 6. Revising to Build Transitions

When writing an early draft, very often you are most concerned simply with getting your main ideas down on paper. Even though you may start with a fairly clear sense of how each idea relates to the others in your text, the bridges that connect them often remain implicit. As you revise in later stages, you need to make the relation between ideas explicit for the reader. One effective way to build links between ideas is to use TRANSITIONS. A transition is a word or phrase that signals the relation between two pieces of information (e.g., two clauses, two sentences, two paragraphs, or even two longer units such as sections of a report).

Types of relations that transitions can show

There are six common types of relations between ideas in a text. First, you may be ADDING a fact or argument to an earlier fact or argument. For example, the second sentence in (1) adds a fact about the class discussed in the first sentence.

(1) Projects in my technical writing classes included copyediting reports and creating a company newsletter. *In addition,* I gained experience with Power Point.

Second, you may be showing a CAUSE-EFFECT relationship. For example, the first sentence in (2) describes an action, while the second sentence describes its consequences.

(2) More than 80 percent of elder abuse cases go unreported. *As a result*, there are relatively few studies of the problem.

Third, you may be showing a CHRONOLOGICAL relation between ideas. For example, passage (3) identifies steps in a procedure.

(3) *After* you obtain a claims form from the Claims Office, complete Items 3 and 4. *Then* return the form to the Claims Office.

Fourth, you may be ILLUSTRATING a general claim made in another part of the text. For example, passage (4) starts with a general claim and then illustrates the claim.

(4) Some observers fear that deregulation of the airline industry has led to decreased safety. To conserve labor costs, *for example*, the airlines sometimes reduce aircraft inspections.

Fifth, you may be COMPARING ideas. For example, passage (5) discusses two similar processes.

(5) The redial feature on a phone lets you access a series of numbers by pushing a single button. *In the same way*, the macro feature in a computer program lets you access a chain of commands by entering just one command.

Finally, you may be CONTRASTING ideas. For example, passage (6) describes a disadvantage and an advantage of a computerized inventory system.

(6) *On the one hand*, computerizing our inventory would require a considerable initial outlay of time and money. *On the other hand*, the computerized system should increase our efficiency and result in fewer back orders to customers.

Types of transitions

Three classes of words are commonly used to link or provide transitions between ideas: COORDINATING CONJUNCTIONS, CONJUNCTIVE ADVERBS, and SUBORDINATING CONJUNCTIONS. The table below provides examples of each type of transition.

	Use to coordinate equal ideas		Use to subordinate one idea to another
	COORDINATING CONJUNCTION	CONJUNCTIVE ADVERB	SUBORDINATING CONJUNCTION
ADDING	*and*	*also, furthermore, in addition, moreover*	
CAUSE-EFFECT	*for, so*	*as a result, because of this, consequently, therefore, thus*	*because, if, since*
CHRONOLOGICAL	*and, then*	*first, second, third. . ., in closing, in conclusion, in short, in summary, next*	*after, as long as, as soon as, before, once, until, when*
ILLUSTRATING		*as an example, for example, for instance, in other words, in particular, specifically*	
COMPARING	*and*	*along the same lines, in the same way, likewise, similarly*	*as, just as*
CONTRASTING	*but, or, yet*	*however, instead, nevertheless, on the one hand, on the other hand, rather, unfortunately*	*although, despite the fact that, even though, in spite of the fact that, though*

The following discussion looks more at each type of transition: its use, placement, and punctuation.

Coordinating conjunctions. Coordinating conjunctions are probably the most common way to link ideas. These conjunctions should be used to join items to which you want to give equal emphasis. One common use for coordinating conjunctions is to join two INDEPENDENT

CLAUSES to make a COMPOUND SENTENCE. (For a review of how to distinguish an independent clause from a DEPENDENT CLAUSE, see Chapter 14, "Editing Sentence Fragments.") Example (7) illustrates this function.

(7)   The department head uses the annual reports to assign merit raises, *and* the dean uses the reports to evaluate the department head's leadership ability.

Note that a compound sentence is punctuated with a comma after the first independent clause.

Because there are relatively few coordinating conjunctions, each one covers a rather broad range of meaning. In particular, *and* functions as a generic link between clauses. For example, in the sentence above, *and* could signal either a temporal relation (more clearly expressed by *Then the dean . . .*) or a comparative relation (more clearly expressed by *In the same way, the dean . . .*). You can often achieve more precision and variety of expression by replacing a coordinating conjunction with a conjunctive adverb or a subordinating conjunction.

Conjunctive adverbs. A conjunctive adverb also joins two units of equal importance. In contrast with the limited number of coordinating conjunctions, however, numerous conjunctive adverbs are available. Consequently, using conjunctive adverbs as you revise will enable you to introduce greater precision and variety.

A conjunctive adverb always relates two independent clauses, the second of which contains the adverb. The two clauses may be separated by either a semicolon, as in (8a), or a period.

(8a) The committee makes recommendations about funding; *however*, the director makes the final decisions.

Conjunctive adverbs can appear in three different places in a clause: at the beginning, in the middle (usually following the subject or an introductory phrase), or at the end. Placing the conjunctive adverb at or near the beginning of its clause is usually best, since the transition can then guide the reader's interpretation of what follows. The conjunctive adverb is separated from the rest of its clause by one or two commas, depending on position, as in (8b)-(8e).

(8b) *However*, the director makes the final decisions.
(8c) The director, *however*, makes the final decisions.
(8d) In the end, *however*, the director makes the final decisions.
(8e) The director makes the final decisions, *however*.

Subordinating conjunctions. A subordinating conjunction links two clauses that the writer does not want to emphasize equally. A subordinating conjunction should be used to introduce a DEPENDENT CLAUSE, the clause that you want to de-emphasize. The dependent clause must then be linked to an independent clause. The resultant sentence, consisting of a dependent clause and an independent clause, is called a COMPLEX SENTENCE.

The following sentence illustrates a common pattern for complex sentences, in which the subordinate clause opens the sentence, and the independent clause closes it.

(9a) *Even though* sand colic is a common disorder in areas of loose, sandy soils, little information exists about its clinical signs, treatment, and prognosis.

This example shows the main pattern for punctuating complex sentences: use a comma between the subordinate clause and the independent clause.

The order of the independent and subordinate clauses can also be varied:

(9b) Little information exists about the clinical signs, treatment, and prognosis of sand colic, *even though* it is a common disorder in areas of loose, sandy soils.

While subordinate clauses offer this flexibility in placement, it is often clearer to keep the subordinate clause at the beginning of the sentence. This way, the transition (e.g., *even though*) directs the reader's interpretation of the material that follows.

Standard written English does not allow the subordinate clause to be detached from an independent clause. The italicized material in the following sentence is a sentence fragment.

More than 80 percent of elder abuse cases go unreported. *Because there are relatively few studies of the problem.*

See Chapter 14, "Editing Sentence Fragments" for more practice in identifying and correcting this pattern.

How transitions help readers

As these examples show, a transition simply helps the reader to see a relationship between ideas. By doing this, transitions reinforce organization because they help the reader follow your idea. They are like road signs: when your reader reaches an "intersection" in your text (e.g., the end of one sentence and the beginning of another), the transition helps the reader to interpret what follows in precisely the way you want. As the writer, you may think the direction is obvious without a signal, but remember that you've traveled the terrain several times before, while thinking about and drafting your text. The reader, in contrast, is making the trip for the first time.

Applications

1.   For each passage below, identify the relation that exists between the sentences (adding, cause-effect, chronological, illustrating, comparing, or contrasting). Then rewrite each passage using a conjunctive adverb to clarify the relation for the reader.

> EXAMPLE: Some software systems have been developed specifically for college bookstores. Missouri Book Service offers a system called *Text-Aid*.
>
> ANALYSIS: Illustrating. Some software systems have been developed specifically for college bookstores. For example, Missouri Book Service offers a system call *Text-Aid*.

a.   WCLP, a company which employs several thousand people, does not have a standardized personal computer system. WCLP has three types of computers: an IBM personal computer, IBM "clones," and a dedicated word-processing station called the "4000."

b.   The idea of being self-insured would seem tempting to a company looking at cutting costs. The risks involved in such a decision are quite severe.

c.   The system we select must not require constant maintenance by outside repair personnel. It should be durable and easy to repair.

2.   For each passage below, identify the relation that exists between the sentences (adding, cause-effect, chronological, illustrating, comparing, or contrasting). Then rewrite each passage using a subordinating conjunction to clarify the relation for the reader.

a.   C-MOS chips are slightly more expensive initially than TTL's. Their use greatly reduces your overall system cost in the long run.

b.   The Seco connectors that you ordered are currently out of stock. We will gladly supply you with Wirecom connectors at the same price.

c.   One reviewer has not responded. I will try to contact the reviewer one more time. I will replace that reviewer with someone else.

d.   I have experience working in the cash office of a busy retail store. I believe I am a good candidate for your internship.

3.   For each passage below, identify the relation that exists between the sentences (adding, cause-effect, chronological, illustrating, comparing, or contrasting). Then rewrite each passage twice, using a different transition in each revision.   You may need to rearrange the order of the clauses in some passages.

a.   A stock split is a good indication that the price of a stock has risen and is doing well. A stock split does not increase a shareholder's relative wealth.

b.   Rising energy costs and declining water tables have increased the cost of irrigating today's farmlands. The efficient use of irrigation water has become an important factor in economically successful farming.

c.   Many people make investment decisions without adequate information.   Most of these people do not know how to acquire such information.

d.   The methods of determining water requirements vary. Their results are all basically the same.

4.   For each passage below, identify the relation that exists between the sentences (adding, cause-effect, chronological, illustrating, comparing, or contrasting). Then rewrite the passage using conjunctive adverbs, subordinating conjunctions, or a combination of these two transitional devices. Some passages will reflect more than one type of relationship.

a.   Calculate the contribution of the forced convection with the Dittus-Boelter equation. Calculate the boiling heat transfer with the Borishansky-Minchendo pool boiling correlation. Add these two together with the Rohsenow superposition expression.

b.   Neuroleptic drugs are widely used to treat patients with schizophrenia. These drugs, which block dopamine receptors, are not successful against all schizophrenia symptomology and may produce side effects such as movement disorders. Taking patients off neuroleptics can trigger the return of schizophrenia in an intensified form.

c.   A positive move for Marks & Spencer would be to expand into Canada and into the U.S. These two countries are very large and could bring in enormous profits. The tax rates are much lower in these countries. Marks & Spencer will probably have to revise many of their operating philosophies to fit these new markets. Adding dressing rooms to the stores in the U.S. and Canada would be required to fit these cultures.

d.   Since structures have been built of concrete, a way to transport the concrete from the mixer to the form has been needed. In the 1950s and 1960s, a wheelbarrow and the Georgia buggy, a motorized wheelbarrow, were used for transportation of concrete. The wheelbarrows took many men and too much time. Today, they are only used on small jobs where very little concrete is needed. The modern solutions to the old transportation problems are the concrete pump and crane. Both have advantages and disadvantages to their use.

e.   ABC Industries has a number of possible alternative courses of action to follow. It could finance expansion by selling large debt securities. This would allow ABC to write off interest expense against income. It could increase advertising to increase sales to absorb some of the effects of the high income tax. This idea probably wouldn't work well. The company would need large increases in unit sales due to the low profit margin on its products. To increase internal control, ABC could require separation of duties for the employees and make use of serially numbered invoices. Although this is expensive to implement, the effective operation of the subordinates will be enhanced.

5.   Revise Appendix 9 ("Executive Summary on 'ProWear Shoes'") for better transitions.

6.   Revise Appendix 11 ("Report on 'Security Methods for Ashley's Clothing Store'") for better transitions.

7.   Revise Appendix 12 ("A Discussion of Prenatal Exposure to Cocaine") for better transitions.

Chapter 7. Revising Format to Unify Text

FORMAT refers to the arrangement of the text on the page (LAYOUT) and visual features of the print (TYPOGRAPHY). Format can unify a text by keeping certain visual features of the text constant as other elements change. This visual continuity provides a context against which the changing information stands out, making it easier for the reader to follow the organization of the text. As an example, try locating the letter "O" in Figures 1 and 2.

```
A E D C R          X X X X X
V X Q B J          X X X X X
T P M O G          X X X O X
Y H I S U          X X X X X
```

Figure 1   Figure 2

Figure 1 has no unified background, so the O does not stand out in contrast. In Figure 2, on the other hand, the repeated element (the X's) focuses your attention on the contrasting element (the O). Similar effects can be achieved by repeating various elements of format--for example, using the same typeface for major headings.

This chapter describes some basic strategies for formatting a document on 8½" x 11" paper, the standard size for letters, memos, and reports in the U.S.. It also discusses some basic typographical effects, all of which can be achieved using a good word-processing program (e.g., MS Word or WordPerfect) and a laser or inkjet printer.

Creating an effective page layout

A well-designed layout is internally consistent, attractive, and consistent with standard practice in your field. As such, layout has both a functional aspect — to make your document easy to read and its structure easy to follow, and an aesthetic aspect — to make your document more attractive to the reader and to reflect positively on you as the writer. The following guidelines reflect standard practice in many companies and organizations, although of course your organization may follow a style guide that differs from these suggestions. (When in doubt about how to lay out a document, try to find an example of a similar, recently produced document and examine its design.)

Margins. Allow at least a 1" margin all around your text. Place page numbers within the margin (e.g., in the upper right or bottom center). The top and bottom margins can also be used for HEADERS or FOOTERS, material that appears on every page (e.g., a chapter title).

Line spacing. More final drafts of technical and professional documents should be single-spaced. (Exception: Manuscripts being submitted for publication as books or articles are generally double-spaced.) In a single-spaced text, leave extra space between paragraphs and before and after headings. This extra space will help the reader to follow your organization.

Beware of "widows" and "orphans" in your text — single lines of type within a paragraph that get stranded at the bottom or top of a page. Most word-processing programs allow you to format for "Widow and Orphan Protection" so that these stranded lines are grouped with the rest of the paragraph.

White space. In addition to margins and line spacing, white space can be used in other ways to improve the readability and appearance of a text. White space serves to set off material from the surrounding text, highlight it, and define it as a unit. Using white space around a vertical list is especially effective for displaying detailed information, since it visually separates the details for the reader. For example, an informal table in a memo, a list of materials required for a procedure, or a list of job responsibilities on a résumé can all be displayed effectively by using white space.

Indentation. Indentation, which is actually a special use of white space, also improves the readability of documents by helping the reader recognize the beginning of paragraphs. (If you are following a format that does indent paragraphs, you will need to leave an extra line of white space between paragraphs.) Indentation can also be used to set of a block of text that serves a special function, such as a long quotation, a list of materials, or a warning.

Justification. JUSTIFIED text is aligned along a vertical margin. Prose texts are normally either FULL-JUSTIFIED (aligned at both the left and right margins) or LEFT JUSTIFIED (aligned only at the left margin, with a RAGGED RIGHT margin). In general, full justification creates a more formal appearance. However, full justification can create distracting gaps between words, especially with a font has nonproportional spacing (as do Courier and Letter Gothic). This problem occurs in example (1a).

(1a) FULL JUSTIFICATION, NONPROPORTIONAL SPACING (COURIER FONT):

```
     Postponing  the  purchase  of  computer  equipment  will
provide   more   time  to  determine  our  future  computing
needs.    This  additional  time  will  also  allow  us  to
consult  with  a  software  specialist  about  customizing
accounting forms.
```

The ragged-right margin in (1b) eliminates these gaps.

(1b) LEFT JUSTIFICATION, NONPROPORTIONAL SPACING (COURIER):

```
     Postponing the purchase of computer equipment will
provide more time to determine our future computing
needs.  This additional time will also allow us to
consult with a software specialist about customizing
accounting forms.
```

Color and shading. Color and shading can be used to draw attention to elements in your document such as headings, graphics, and inset material. Be careful, though, not to go overboard in your use of colors. Also, since colors and shading do not always photocopy clearly, think about how your document will be used and reproduced before you introduce these components.

Using typography for emphasis and unity

After designing an effective page layout for your document, you can further use typography to emphasize and unify certain elements (e.g., headings). This section discusses some common typographical effects and their uses.

A word of caution: the capabilities of word processors may tempt you to "go overboard" mixing typefaces and other effects. The result can be a text that looks cluttered and chaotic instead of unified and professional. When in doubt, opt for fewer variations.

Typeface. TYPEFACE refers to the overall design of a print alphabet. Typefaces are either SERIF or SANS SERIF. A serif typeface has tiny "feet" at the end of some letters, whereas a sans serif typeface does not: This is a serif typeface. This is a sans serif typeface. Serif typefaces tend to give a more traditional flavor than sans serif typefaces. In general, a serif typeface is a better choice for extended prose in printed elements (such as the body of a text). Sans serif typefaces can be used for shorter elements such as headings. On the other hand, serif typefaces are a better choice for both body and headings in online documents.

Type size. Type size is measured in POINTS of 1/72". The average typewriter is limited to one type size. On the other hand, a laser printer may allow you to vary type size from, say, 6 points to 100 points. The following numerals illustrate a range of point sizes in Times Roman.

6 10 12 16 24 36

One type size (usually 10 to 12 points) is more than adequate for everyday memos, letters, and short reports on 8½" x 11" paper. (There are exceptions, of course; for example, a larger type size is better for instructions that will be read at a distance.) In longer manuscripts, you may want to add a larger type size to unify major headings such as chapter titles.

Centering. Centering is very easy on most word processors. Because readers normally expect text to begin at the lefthand margin, centering has limited uses within most documents. However, the items on a title page are commonly centered. Another use for centering is to set off first-level (i.e., major) headings, such as the chapter headings in this textbook.

Capitalization. In addition to capitalized, or upper-case, letters, most word-processing programs also allow for small capitals: THIS TEXT IS IN NORMAL CAPITALS; THIS TEXT IS IN SMALL CAPITALS. Because capitals are hard to read for long stretches of text, they are best reserved for headings or short phrases. One common use for capitals is to designate first-level headings. Capitals can also be used to emphasize a word, phrase, or sentence, as in (2).

(2)   Do NOT use this product within 5 feet of an open flame.

Small capitals can be used to introduce a technical term, as in (3). (When you use this convention, the reader will expect a definition to follow in the text or in a glossary.)

(3)   A DEBIT is the recording of debt in an account.

Boldface. Printing text in boldface creates thicker, darker letters. One function of boldface is to unify and emphasize headings (as in this textbook). Second, boldface can be used to emphasize a word, phrase, or sentence, as in (4).

(4)   Do **not** use this product within 5 feet of an open flame.

Third, boldface can be used to identify a technical term the first time it is introduced, as in (5).

(5)   A **debit** is the recording of debt in an account.

To avoid confusing your reader, do not mix these last two uses of boldface in the same text. For example, if you are using boldface to designate technical terms, use another typographical effect (such as capitalization) for emphasis.

Underlining. Underlining serves similar functions to boldfacing. First, it can be used to unify headings. Example (6) uses capitals for first-level headings and underlining for second-level headings.

(6)   FIRST-LEVEL HEADING

    XXXXXXXX XXXXXXX XXXXXXXX XXXXXXX XXXXXXXXXX XXXXXXXXX XXXXXXXXXX XXXXXXXXXXXXXX XXXXXXXXXXXXXXXX XXXXXXXX XXXXXXXX XXXXXX.

Second-level Heading

    XXXXXXXX XXXXXXX XXXXXXXX XXXXXXX XXXXXXXXXX XXXXXXXXX XXXXXXXXXX XXXXXXXXXXXXXX XXXXXXXXXXXXXXXX XXXXXXXX XXXXXXXX XXXXXX.

Second, underlining can be used to emphasize a word, phrase, or sentence, as in (7).

(7)   Do not use this product within 5 feet of an open flame.

Third, underlining can be used to identify a foreign word or the first use of a technical term, as in (8).

(8)   A debit is the recording of debt in an account.

Do not use underlining for both emphasis and foreign or technical terms. For example, if you underline for emphasis, use another device such as italics for technical terms.

Italics. In an ITALIC style, the letters of a typeface lean slightly to the right: *This is italicized Times Roman;* `This is italicized Courier`. Although italics alone are not recommended for first- or second-level headings, they can be used for lower-level headings such as paragraph headings. Italics are used for the third-level heading in example (9).

(9)   FIRST-LEVEL HEADING

    xxxxxxxx xxxxxxx xxxxxxxx xxxxxxx xxxxxxxxxx xxxxxxxxx xxxxxxxxxx xxxxxxxxxxxxxx xxxxxxxxxxxxxxxx xxxxxxxx xxxxxxxx xxxxxx.

    <u>Second-level Heading</u>

    xxxxxxxx xxxxxxx xxxxxxxx xxxxxxx xxxxxxxxxx xxxxxxxxx xxxxxxxxxx xxxxxxxxxxxxxx xxxxxxxxxxxxxxxx xxxxxxxx xxxxxxxx xxxxxx.

    Third-level Heading. xxxxxxx xxxxxxxx xxxxxxx xxxxxxxxxx xxxxxxxxx xxxxxxxxxx xxxxxxxxxxxxxx xxxxxxxxxxxxxxxx xxxxxxxx xxxxxxxx xxxxxx.

Another use for italics is to identify a foreign word (for example, *habeas corpus*) or the first use of a technical term.

Using characters to list or enumerate

Numbers and letters are commonly used to designate items in a series. Numbers include Arabic numerals (1, 2, 3), upper-case Roman numerals (I, II, III), and lower-case Roman numerals (i, ii, iii). Letters can be either upper case (A, B, C) or lower case (a, b, c). Numbers and letters alternate in a traditional outline, shown in example (10).

(10) I.   First-level heading
     A.   Second-level heading
     B.   Second-level heading
       1. Third-level heading
       2. Third-level heading
    II.   First-level heading

The decimal number system, illustrated in (11), is often used in technical discourse.

(11) 1.   First-level heading
     1.1. Second-level heading
     1.2. Second-level heading
       1.2.1. Third-level heading
       1.2.2. Third-level heading
    2.   First-level heading

Numbers or letters may also be integrated into a text. The most common items for this use are lower-case letters or Arabic numerals, as in (12).

(12) Skin cancer falls into three major categories: (1) melanoma, (2) squamous cell carcinoma, and (3) basal cell carcinoma.

Likewise, prose numbers (*First, Second,* etc.) can be used when a list is integrated into a text, especially to enumerate whole sentences or paragraphs.

Finally, an asterisk (*), dash (--), or bullet (•) can be used to create a stacked (i.e., vertical) list of items on the same level of generality, as in (13). However, these symbols are less effective than numbers or letters for indicating chronology or order of importance.

Skin cancer falls into three major categories:
- melanoma
- squamous cell carcinoma
- basal cell carcinoma

Applications

EXAMPLE: REVISE THE FOLLOWING TEXT SO THAT TYPOGRAPHY IS USED DISTINCTIVELY. The first inflationary model, **constant dollar**, considers only general price-level inflation. The next model, **replacement cost**, considers only specific price changes. The last model, **current cost**, is **claimed** to have the best aspects of both other models.

REVISION: The first inflationary model, **constant dollar**, considers only general price-level inflation. The next model, **replacement cost**, considers only specific price changes. The last model, **current cost**, is *claimed* to have the best aspects of both other models.

1.   Use enumerative devices to format each of the following lists in two different ways: one as a list integrated into the text, and one as a stacked list. Can you state a rationale for preferring one of these formats in each case?

a.   Treatment included bronchodilatros and antimicrobials for COPD; guttural pouch catheterization; povidone iodine lavage; and antimicrobials for the guttural pouch infection.

b.   The design alternatives considered for this project included steel foundation poles with pile footings, drilled piers, mats, spread footings, and an alternative site.

c.   You should get a hammer, Phillips-head screwdriver, two 3/4" bolts, two washers, and two 3/4" nuts before you begin.

2.   Use typography to unify the following passage.

The Return on Sales ratio indicates the profit on each dollar of sales. The Return on Assets ratio indicates how effectively a firm is using its assets. The Return on Equity ratio measures a firm's ability to generate a return for its shareholders.

3.   Revise the following passage so that typography is used distinctively.

The package inset clearly states that the test is intended *for primary isolation only*. Although it supports the growth of *H. influenza*, the test was apparently formulated for primary isolation of *N. gonorrhoeae*.

4. Develop headings for a description of three subsystem tests: the GPI initialization test, the GPI loopback test, and the GPI scratchpad memory test. The description of each test will include the test's objectives, its software and hardware components, any prerequisites for the test, and the data generated by the test. Use distinctive typography to unify each level of heading.

5. Revise each passage below so that it has a more effective format.

    a.   FROM THE "EMPLOYMENT" SECTION OF A RÉSUMÉ:

EMPLOYMENT
    From 1995 to date, I have worked as an accountant at LSU Accounting Services in Baton Rouge, Louisiana. I prepare student payroll, cancel checks, and prepare checks and garnishments for the five campuses of the LSU system. I also supervise three student employees.
    I worked as an account clerk at the LSU Center for Wetland Resources from 1992 to 1995. My duties included purchasing and shipping/receiving, and I processed travel vouchers and maintained equipment inventory records. I also supervised one full-time travel clerk and four student employees.

    b.   FROM A USER'S MANUAL FOR A COMPUTER PROGRAM: Place the backup copy of the hvt.kev program in the primary disk drive. It is recommended that the program be transferred into the directory which contains gwbasic on the hard drive, if you have one; otherwise it can be copied onto the flexible disk which contains gwbasic. When you type hvt.kev, the screen will clear, and a title page will be displayed. Instructions will be given in order to continue to the next page. You will be given a choice of reading the instructions or proceeding with the program by selecting z and then pressing enter.

c.   FROM A PROPOSAL: The following timetable will be followed, upon your approval of this project. During the first two weeks of the project, letters to professional organizations will be sent; software distributors will be required to provide free trial packages of their word processing software; and periodicals will be checked for reviews and advertisements. During the next two weeks, the information received will be analyzed, and a bibliography will be created from the most useful sources. A progress report will be completed in the fifth week. The next two weeks will be used for organization, analysis, and preparing the final report. The total time for this study will be seven weeks.

d.   FROM A MEMO TO EMPLOYEES: Because of tax increases, we are forced to cut back on expenses. After lengthy consideration of the options available, it was decided that we should change the current pension plan slightly. Starting July 1, we will contribute an initial 5% of your salary (vs. 10% currently) to your retirement plan. If you desire, we will then match your personal contribution of 2-1/2%, allowing you to maintain your present 10% pension. In other words, you may choose 10% benefits with a personal contribution of 2-1/2%, or 5% benefits without any contribution. This would mean at your current salary level of $35,000 we would provide you with $1750; you would then have the option of contributing up to $875 from your salary, which we would match for a total of $3500 per year (the current total level).

e.   FROM AN ARTICLE ON MINIMIZING GANG INVOLVEMENT: There is no single solution to gang problems. Some approaches, however, have a far better chance than others to minimize gang involvement (McEvoy 1990). There are many different prevention plans and intervention plans. The majority are targeted at specific groups. The gang anti-violence program is targeted at gang members and at risk youth. It utilizes two approaches to reducing gang violence. First, outreach workers (a group of community members, social workers, and former gang members who work with opposing gangs to declare schools, churches, parks, and community centers "safe zones") in the target community work with the leaders of local gangs to suppress gang activity and negotiate peace. The program also works with at-risk youth to encourage them to stay in school, assist them with academic and family problems, and train them in the skills needed to find and hold a job. Another gang intervention program is aimed at high school males who are already involved with gangs. It deals primarily with the Gangster Disciples, Latin Kinds, Satan Disciples, and Tat Boys. This program tries to help gang members increase their self-esteem and enhance their coping skills through individual counseling and group workshops. Conflict resolution classes aim at improving their anger management skills while demonstrating that there are alternatives to violence. The gang violence reduction program is aimed at hard-core gang youth ages 17-25. Under this program, a team of civilian gang outreach workers maintains regular contact with up to 200 known gang members, providing them with crisis counseling, job and school referrals, and other alternatives. The team also works with police to defuse conflicts among street gangs before they turn lethal. Additional probation officers provide close supervision to known gang members, and a Community Advisory Group consults with the project staff.

6.   Revise Appendix 1 ("Memo to Residents") so that it has a more effective format.

7.   Revise Appendix 4 ("Brian Carter's Résumé") so that it has a more effective format.

8.   Revise Appendix 6 ("Memo Assessing an Employee's Writing"), Part B, so that it has a more effective format.

Chapter 8. Revising for Conciseness

As you revise on the sentence level, you will want to pay special attention to WORDINESS. Wordiness refers to a stylistic problem which results from the overuse of empty words. Revising a wordy text requires stating its message more concisely — in fewer words — without sacrificing important information, detracting from readability, or creating an undesirable tone. While wordiness can take a number of forms, here we focus on three of the most common patterns that lead to this problem. If you learn to look for and revise these patterns, you will go a long way toward becoming a more concise writer.

Overuse of main verb *be*

Saying that "X is" makes about the most general claim possible about X. Therefore, changing main verb *be* to another verb creates a more specific, informative claim about your subject.

Identifying main verb *be*. The verb *be* has the following forms:

be	*am*
being	*are*
been	*was*
is	*were*

The MAIN VERB in a clause is always the rightmost verb in the verb phrase; any verbs before the main verb are AUXILIARY VERBS. For example, in the sentence *Management is being forthright*, *being* is the main verb and *is* is an auxiliary verb. In the sentence *Management should have been forthright*, *been* is the main verb, while *should* and *have* are auxiliary verbs.

Locating a more specific verb. Main verb *be* is generally followed by a noun, an adjective, or a prepositional phrase (a phrase starting with a word such as *of, on, in, to, by, for, from, with* or *about*). Try looking for a "hidden" verb in the element following the *be*-form. Examples (1)-(3) illustrate this strategy.

(1a) WORDY: This report *is an analysis of* the problem.
      STRATEGY: Try converting the noun *analysis* into a main verb.
(1b) REVISED: The report *analyzes* the problem.

(2a) WORDY: The administration *is supportive of* our program.
      STRATEGY: Try converting the adjective *supportive* into a main verb.
(2b) REVISED: The administration *supports* our program.

(3a) WORDY: This information could *be of benefit to* the students.
      STRATEGY: Try converting the prepositional phrase *of benefit* into a main verb.
(3b) REVISED: This information could *benefit* (or *help*) the students.

Expletive structures

An EXPLETIVE sentence begins with *There* or *It*, followed by a form of *be*. (Expletive openings include *There is*, *It was*, *There could be*, and *It should have been.*) Thus, expletive openings constitute a special case of main verb *be*, along with *There* or *It* as the subject. Note that an expletive structure places relatively uninformative elements — *There/It* and *be* — in two key positions, subject and main verb.

To revise an expletive structure, look for another noun and verb following the expletive. Then make these elements your subject and main verb. Examples (4)-(6) use this approach.

(4a) WORDY: *It is* necessary for us to find a solution to this problem.
     STRATEGY: Try moving *us* into subject position (*we*). Try converting *necessary*, *find*, or *solution* to a new main verb.
(4b) REVISED: *We need* to find a solution to this problem.
                  We must find a solution to this problem.
                  We must solve this problem.

(5a) WORDY: *It is* the purpose of this report to identify the best laser printer.
     STRATEGY: Try making *report* the subject and *identify* the main verb.
(5b) REVISED: *This report identifies* the best laser printer.

(6a) WORDY: *There will be* a demonstration of the new system today.
     STRATEGY: Use *demonstration* as the basis for a new main verb. You will have to create a subject.
(6b) REVISED: *A company representative will demonstrate* the new system today.
                  Come to a demonstration of the new system today.

Nominals

A NOMINAL is a noun derived from a verb, usually by adding *-ance*, *-ence*, *-ion*, or *-ment* to the verb. For example, the noun *maintenance* derives from the verb *maintain*; the noun *achievement* derives from the verb *achieve*. You can achieve a more concise style by converting the nominalized form back to a verb, as shown in (7).

(7a) WORDY: His memo contains *a recommendation* that we purchase a LaserJet.
     STRATEGY: Use *recommendation* as the basis for a new main verb.
(7b) REVISED: His memo *recommends* that we purchase a LaserJet.

Often, a nominal is followed by a prepositional phrase, which can be converted to the object of the new main verb. Examples (8)-(9) illustrate this pattern.

(8a) WORDY: Who is responsible for *the supervision of the temporary staff*?

     STRATEGY: Use *supervision* as the basis for a new main verb; use *the temporary staff* as its direct object.

(8b) REVISED: Who *supervises the temporary staff*?

(9a) WORDY: The article includes *a discussion of methods* for *the retention of employees*.

     STRATEGY: Use *discussion* and *retention* as the basis for new main verbs. Use *methods* and *employees* as direct objects.

(9b) REVISED: The article *discusses methods* for *retaining employees*.
                    The article *discusses ways to keep employees*.
                    The article *explains how to retain employees*.

A caveat about conciseness

As mentioned earlier, be careful not to "throw the baby out with the bath water" by eliminating significant information or creating an undesirably abrupt tone when you revise for conciseness. For example, consider the sentence *The article includes a discussion of methods for the retention of employees*. You can shorten this sentence by saying *This article discusses employee retention*. However, this revision leaves out important information about the article: namely, that it focuses on *methods* of retaining employees, rather than, say, on the importance of retaining employees or employee retention rates in different industries. Thus, a better revision of the original sentence would be *This article discusses methods of retaining employees*.

Tone can also be damaged by cutting out "buffer" material, especially in negative messages. For example, suppose you work in the admissions office of a university and you receive a late application for a scholarship. You might include a sentence in your reply like *The application deadline passed last week on February 1*. Certainly you could make this sentence more concise: for example, *You missed the February 1 deadline*. Obviously, though, what this revision gains in conciseness it loses in goodwill. Chapter 12, "Revising to Improve Tone," discusses strategies for maintaining a courteous tone.

In closing, not all instances of main verb *be*, expletive openings, and nominals can, or should, be eliminated. However, relying too heavily on such constructions often leads to wordiness, so keep these patterns in mind as you revise.

Applications

Revise each sentence or passage for a more concise style.

> EXAMPLE: It is the intention of this report to find the best possible word processor at a reasonable cost.
> REVISION: This report identifies the most cost-effective word processor available.

1.   My internship experience has strengthened my communication skills to help relate information to my peers and superiors.

2.   There are very few people who would buy a security if they knew that they would have to keep it for the rest of their lives.

3.   The first part of this report provides a definition of a portfolio and how it can be beneficial to the investment decisions of Northern Telecom.

4.   It is believed that the responses of the 15 institutions we have contacted so far will yield sufficient data to complete our analysis.

5.   The basic approach of this lease-buy analysis involves the computation of the present value costs associated with leasing and borrowing to buy.

6.   Some investments which are common for the average person would be the purchasing of a new home or vacation site.

7.   The maintenance of the customer's goodwill is the most basic issue that this company has to face.

8.   The clear problem that has arisen is that the benefit of the air contaminate model is often maligned by having to abort launches, perhaps unnecessarily.

9.   As a medical assistant/receptionist, my primary responsibilities included the completion and processing of insurance claim forms. I assisted Dr. Brown during minor surgeries, preparing the patient and examination room for surgeries. I also assisted in the manufacturing, fitting, and dispensing of orthotic devices.

10. As recipients of third-party payments, psychiatric hospitals have received increasing demands for verification of pathology and objective measurements of progress in treatments. This is a proposal for an exploration into the implementation of measurements of wellness into the treatment milieu at Parkway Psychiatric Hospital. Approximately one month will be required to complete the study, which will include (1) discussion of statistical measurements of wellness and (2) recommendations for implementation of the measurement instruments into the hospital's assessment package.

11. There is a large amount of uncertainty inherent to dispersion modeling, and this uncertainty in the predicted toxic corridors has given rise to what may be overly conservative decisions of when to abort and when to proceed with a mission.

12. Revise Appendix 2 ("Chris Applegate's Job Application Letter") for a more concise style.

13. Revise Appendix 7 ("Clinical Guidelines for Electrosurgery") for a more concise style.

14. Revise Appendix 10 ("Report on 'Improving Employee Training at Becker Foods'") for a more concise style.

Chapter 9. Revising for Parallel Structure

Professional writing often uses series and lists that break an entity into smaller parts so that the reader can understand it more clearly. For example, you may need to explain the steps in a procedure; describe the features of a product; or identify alternative solutions to a problem and list criteria for making a decision. In some cases, it may take only a sentence to list a series. In other cases, you may want to use the list to construct section headings and subheadings that reflect the structure of a larger report.

In all of these instances, the series or list will be clearer if you use PARALLEL STRUCTURE. This means that the elements in a series have the same grammatical form (i.e., the same part of speech, same suffix, etc.). When the elements do not have the same grammatical form, they are said to display FAULTY PARALLELISM. This chapter reviews some common sources of faulty parallelism and illustrates ways to revise for parallel structure.

General revision strategies

When revising a draft that contains a series, start by isolating the series if it is distributed throughout the text. For example, suppose you want to make sure that your report headings have parallel structure. As you go through the report, jot down the headings on a piece of paper. It's much easier to compare their structure if you can see them all at once.

Second, use a stacked list (i.e., a vertical format) to display the items being revised, especially when dealing with a complex series. For example, (1a) appears to contain a series of parallel verb phrases.

(1a) The software may be programmed to print checks with each employee's name and weekly salary, keep an up-to-date wage total, a summary of tax deductions, and file this information in individual employee accounts.

However, converting the series to a stacked list, as in (1b), reveals its faulty parallelism.

(1b) • print checks with each employee's name and weekly salary
      • keep an up-to-date wage total
      • a summary of tax deductions
      • file this information in individual employee accounts

The first, second, and fourth items are parallel verb phrases. The third item, though, is a noun phrase and therefore not parallel. This problem is corrected in (1c) by joining the third item to the second item, as an object of the verb *keep*.

(1c) • keep an up-to-date wage total and a summary of tax deductions

The series in (1d) now has parallel structure, since it now consists of three verb phrases.

(1d) The software may be programmed to print checks with each employee's name and weekly salary, keep an up-to-date wage total and a summary of tax deductions, and file this information in individual employee accounts.

Faulty parallelism caused by mixing tenses

When describing a series of events that occurred at the same time relative to the document you are writing, use a consistent tense to report the events. For example, (2a) describes two experiments conducted before the report was written.

(2a) FAULTY: Lorist et al. found that caffeine affects the dynamic state of participants after 50-250 mg and can last up to five hours. Jarris' research also shows that caffeine . . .

The verb in the first sentence is past tense (i.e., *found*), but the verb in the second sentence is present tense (*shows*). This passage should be revised by using one tense for both verbs in the series, as in (2b).

(2b) PARALLEL: Lorist et al. found that caffeine affects the dynamic state of participants after 50-250 mg and can last up to five hours. Jarris' research also showed that caffeine . . .

Introductions to long documents are usually written in present tense. Passage (3a) contains four verb phrases referring to a report and its sections. Note that the first two are present tense (*analyzes, reviews*), while the second two are future (*will describe, will identify*).

(3a) FAULTY: This report analyzes career opportunities at the entry level for a Certified Public Accountant (CPA). The first section reviews the requirements for acquiring and maintaining a CPA license. The second section will describe the skills needed to succeed as a CPA. The third section will identify growth patterns in specialty areas.

The revision in (3b) changes all of the verbs to present tense.

(3b) PARALLEL: This report analyzes career opportunities at the entry level for a Certified Public Accountant (CPA). The first section reviews the requirements for acquiring and maintaining a CPA license. The second section describes the skills needed to succeed as a CPA. The third section identifies growth patterns in specialty areas.

Faulty parallelism caused by mixing verb voices

Another common problem is alternating between active and passive voice in a series, as in (4a).

(4a) FAULTY: This report describes existing conditions, and future wastewater flows are also predicted.

The first verb phrase is in the active voice (*describes*), while the second is in the passive voice (*are also predicted*). The revision in (4b) uses one voice for both verb phrases.

(4b) PARALLEL: This report describes existing conditions and predicts future wastewater flows.

For further review in distinguishing active and passive voice, see Chapter 10, "Revising Active and Passive Voice."

Faulty parallelism caused by mixing different types of verbals

A VERBAL is a verb used in place of a noun phrase or a modifier. The following table identifies the three types of verbals: INFINITIVES, GERUNDS, and PARTICIPLES.

	INFINITIVE	GERUND	PARTICIPLE
TYPICAL FORM	*to* + verb	verb + *ing*	verb + *ing* verb + *ed* verb + *en* verb + other past participle affix
TYPICAL FUNCTION	noun phrase (subject, object, or complement)	noun phrase (subject, object, or complement)	adjective (noun modifier or complement)
EXAMPLE	The next step is *to freeze* the specimens.	*Freezing* the specimens is the next step.	Our experiment involved three *frozen* specimens.

Verbals in a series should all be of the same type. For example, (5a) mixes two infinitives with a gerund.

(5a) FAULTY: The purpose of this report is to describe existing conditions, to predict problems, and estimating the cost of improvements.

For parallel structure, the passage needs one type of verbal. Version (5b) uses infinitives.

(5b) PARALLEL: The purpose of this report is to describe existing conditions, predict future problems, and estimate the cost of improvements.

Faulty parallelism caused by mixing verbals and nominals

A NOMINAL is a noun derived from a verb, usually by adding *-ance*, *-ence*, *-ion*, or *-ment* to the verb. For example, *election* is derived from the verb *elect*. Mixing a nominal with a verbal results in faulty parallelism, as in (6a).

(6a) FAULTY: The purpose of this report is to describe existing conditions and the prediction of future wastewater flows.

The complements in this sentence are not parallel because they do not have the same grammatical form. This passage should be revised by using either a verbal or a nominal for both complements. Version (6b) uses verbals.

(6b) PARALLEL: The purpose of this report is to describe existing conditions and to predict future wastewater flows.

Faulty parallelism caused by mixing different types of noun clauses

Several types of clauses can function in noun positions (subject, object, or complement) in a larger sentence. Generally defined, a CLAUSE is a series of words containing a subject and a predicate (all sentences are clauses, but not all clauses are sentences). Two common types of noun clauses are WH-CLAUSES (introduced by *who*, *what*, *when*, *where*, *whether*, or *how*) and THAT-CLAUSES. These two clause types should not be mixed in the same series; compare (7a) and (7b).

(7a) FAULTY: He reported on what the existing conditions are and that they will have an environmental impact.
(7b) PARALLEL: He reported that the existing conditions are unstable and that they will have an environmental impact.

Faulty parallelism caused by mixing clauses or phrases with sentences

A final source of faulty parallelism is mixing complete sentences (i.e., independent clauses) with noun clauses or phrases. This problem occurs in (8a).

(8a) FAULTY: Several government actions limited the company's financial capabilities:
1) Raising the corporate tax rate from 40% to 52%.
2) The gross equivalent of ordinary dividends declared in 1996 could not exceed those of the previous year by more than 5%.
3) Forcing all retailers to reduce gross margins by 10% in 1997.

This list contains two verb phrases (items 1 and 3) and one sentence (item 2). Since item 2 is too complex to reduce to a phrase, the best strategy here is to change items 1 and 3 to sentences, as in (8b).

(8b) PARALLEL: Several government actions limited the company's financial capabilities:
1) The corporate tax rate increased from 40% to 52%.
2) The gross equivalent of ordinary dividends declared in 1996 could not exceed those of the previous year by more than 5%.
3) All retailers had to reduce gross margins by 10% in 1997.

Applications

The passages below have faulty parallelism. Revise each passage so that it has parallel structure. You may want to begin by isolating the items in each series and displaying them in a stacked list.

> EXAMPLE: Potential employees who show personal integrity, a commitment to service, and that they can ski are prime candidates for employment at Head Ski.
> REVISION: Potential employees who show personal integrity, a commitment to service, and an ability to ski are prime candidates for employment at Head Ski.

1. The manager at DuPont stated that about 60% of engineers hired had little or no experience. Similarly, the director of personnel at Dow Chemical states that only 20% of engineers hired were experienced.

2. This report presents a description of asphalt types. Then an overview of research on the durability of these types is presented. In conclusion, I will present the reasons for choosing Type 2 asphalt for the Burt Co. project.

3. The proposed study will provide information on possible causes and prevention of embryonic waste. We will also investigate whether HCG enhances P4 production. Finally, the ability of HCG to counteract a hypothesized negative effect of PGF on the developing corpeus luteum will be investigated.

4. For those with a background in early psychological and philosophical theories of the mind, and are looking to discover more recent advances in theories of the mind, this book is a good choice.

5.   The software must do the following:
     (a)   handling of both incoming and outgoing books
     (b)   differentiate between textbooks and general books
     (c)   give credit for book returns
     (d)   printing customized forms
     (e)   generate operation reports

6.   My education and experience show that I have the most important skills an HNB employee can have: to work well with employees and customers, and understanding and following bank policies.

7.   All of our employees receive automatic step raises, bonus plans, and are included in our profit-sharing system.

8.   Many factors should be considered before a final decision to enter the restaurant business can be made. These include the amount of capital you have or can obtain, the amount of risk you are willing to incur, how much independence you want, and do you really have what it takes to make a new restaurant successful.

9.   Bird and Ross (1994) found that relative to paid work, housework provides more autonomy. However, they also found housework to be more routine, less intrinsically gratifying, and fewer extrinsic rewards.

10. Advantages to opening retail markets are:
    1. allows us to compete head-on with retail outlets, and
    2. may increase brand recognition.
    Disadvantages to opening retail markets are:
    1. operating costs will be increased, and
    2. some catalog sales may be lost.

11. FROM A JOB DESCRIPTION ON A RÉSUMÉ:
    Office Assistant in Financial Aid
    -used computer to help calculate students' aid award amounts
    -frequent use of Lotus
    -worked with confidential informational needs
    -various clerical activities

12. The proposed study consists of the following six steps.
    1. Analyzing Box Pack's debt structure and financial statements.
    2. Comparing figures and financial ratios with the industry norms.
    3. Analysis of the advantages and disadvantages of each alternative with respect to ownership of the company.
    4. Compute which alternative will be the most cost-efficient for the company.
    5. Analyze the net tax effects of each alternative.
    6. A recommendation about which debt alternative Box Pack should use.

13. FROM A PROPOSAL:
     Schedule
     1.   I plan to research extensive pension fund investments that are commonly used in the first two weeks of study.
     2.   After obtaining a thorough background of investment weights in various possible portfolios, one-half week will be devoted on the possible advantages and disadvantages of these solutions for a company.
     3.   Discussing the optimal solution discovered in Step 1 or 2 of the schedule in the remaining part of the third week.
     4.   Provide expected returns and results to the company.

14. Characteristics of caregivers who are likely to be abusers are: being a problem drinker; abuser of drugs; psychologically impaired; mental or emotional illness; inexperienced at caregiving; economically troubled; abused as a child; faces emotional, social, professional or economic stress; unengaged outside of the home; blames the older person for problems related to the pressures or burdens of providing; has unrealistic expectations; economically dependent; and is hypercritical.

Chapter 10. Revising Active and Passive Voice

One important decision as you revise at the sentence level is whether to use ACTIVE or PASSIVE VOICE in a particular sentence. You may have heard advice like "Use active voice unless passive voice is more appropriate." This is often sound advice because, as we will see, active voice is more direct and concise. In a number of cases, however, passive voice may create a more effective style or tone. This chapter shows how to identify active and passive voice and provides some guidelines about when to use each type of structure.

Identifying active and passive structures

The distinction between active and passive voice is related to the THEMATIC ROLES played by the noun phrases in a sentence. The two main thematic roles in a sentence are AGENT and PATIENT. The agent is the noun that performs the action described by the verb. The patient is the noun affected by the action of the verb. For example, given the nouns *chemist* and *test tubes* and the verb *filled*, we would expect *chemist* to be the agent and *test tubes* to be the patient.

In English, the unmarked, or normal, position for the agent is as the SUBJECT of a sentence: the first noun phrase in the main clause. The unmarked position for the patient is as the DIRECT OBJECT: the noun phrase following the main verb. In an active structure, the agent is in subject position and the patient (if any) is in direct object position. This pattern is illustrated in (1)-(3).

THEMATIC ROLE:	AGENT	ACTION	PATIENT
SYNTACTIC POSITION:	SUBJECT	VERB	DIRECT OBJECT
(1a)	The chemist	filled	the test tubes.
(2a)	Customers	have written	several letters of complaint.
(3a)	We	are conducting	further research.

It is possible to have an active structure without a direct object: for example, *Customers have complained*. However, since a patient is needed to convert a sentence to passive voice, we will focus on sentences with both an agent and a patient.

Deriving a passive structure from the corresponding active involves four changes:

- Moving the patient into subject position.
- Adding a form of *be* immediately before the main verb. (The forms of *be* are *am, are, be, been, being, is, was,* and *were*.)
- Putting the main verb into its past participle form. (This is the form that fits into the slot *I have* _____ or *It has* _____.)
- Moving the agent into a *by*-phrase after the main verb.

These changes are illustrated in the following sentences.

THEMATIC ROLE:	PATIENT	ACTION	AGENT
SYNTACTIC POSITION:	SUBJECT	VERB	OBJECT OF *BY*
(1b)	The test tubes	were filled	by the chemist.
(2b)	Several letters of complaint	have been written	by customers.
(3b)	Further research	is being conducted	by us.

It is further possible to omit the *by*-phrase from a passive structure, thereby creating an AGENTLESS PASSIVE: for example, *The test tubes were filled*; *Further research is being conducted*. You can still identify these as passives by examining the verb phrase: it contains a form of *be*, immediately followed by a past participle form of the main verb.

Note, by the way, that passive voice is not the same as past tense. It is possible to have a passive sentence in the present tense (e.g., *Further research is being conducted*). Likewise, it's possible to have an active sentence in the past tense (e.g., *The chemist filled the test tubes*).

Choosing between active and passive voice

This section reviews several criteria you can apply during revision to decide whether a sentence should be in the active or the passive voice.

Use active voice for a more concise form. If reducing the number of words is your only concern as you revise a particular sentence, use the active voice. You will be able to state the same information in fewer words. This is because the passive voice introduces additional elements — a form of *be* and the preposition *by* — that automatically increase word count.

(2a) ACTIVE (7 WORDS): Customers have written several letters of complaint.
(2b) PASSIVE (9 WORDS): Several letters of complaint have been written by customers.

Of course, you can reduce the word count in the passive version by omitting the agent. However, you are also reducing the amount of information in the sentence.

Use active voice to focus on the agent. Active sentences are more effective when you need to specify the reader or some other entity as the agent of the action. This guideline is especially important if you are writing instructions or assigning tasks. In contrast, an agentless passive is a poor choice in these contexts because it is vague about exactly who the agent is. For example, compare (4a) and (4b).

(4a) AGENTLESS PASSIVE: These letters of complaints should be answered.
(4b) ACTIVE: Our Public Relations Director, Chris Marlowe, should answer these letters of complaint.

The active version assigns responsibility for the action to a specific agent.

Use passive voice to focus on the patient. Sometimes, you may want to de-emphasize an agent's role in performing an action. You may want to focus instead on the patient, the entity affected by the action. Because passive voice rearranges the agent and patient, it allows you to "demote" the agent into a *by*-phrase and "promote" the patient into subject position. For example, compare (5a) and (5b).

(5a) ACTIVE: Women tend to use more "polite" linguistic forms than men do. They are also less likely to interrupt another person.

(5b) PASSIVE: "Polite" linguistic forms tend to be used more by women than by men. In courtroom settings, such forms also tend to be used by speakers of low socioeconomic status.

Note that each passage begins with a different version of the same sentence. The first passage focuses on women as users of language; therefore, the active version is more appropriate because it places the agent (*Women*) in subject position. In contrast, the second passage focuses on how polite forms are used by different groups. The passive voice is more appropriate here, because it places the patient (*"Polite" linguistic forms*) in subject position.

You can also use an agentless passive when the agent's identity is obvious, unknown, or confidential. For example, suppose you are preparing a progress report that contains the passage (6a).

(6a) DRAFT: I am currently collecting information about programs from three manufacturers. I have asked each manufacturer to send a demonstration disk.

Since you, the writer, are obviously the agent, you can recast the second sentence in the passive voice to emphasize the effects of your actions, as in (6b).

(6b) REVISED: I am currently collecting information about programs from three manufacturers. Each manufacturer has also been asked to send a demonstration disk.

Agentless passives are useful if the agent is unknown; for example, *Three staplers were stolen from the supply room.* And in the following sentence, the agentless passive allows the writer to maintain the informant's confidentiality: *We have been informed that you are releasing proprietary information about clients.*

Use passive voice to describe generalizable actions or results. Passive voice is appropriate for describing widespread beliefs or actions: for example, *Today, less rigid plastics are used throughout the automotive industry for bumpers and moldings.*

One interesting pattern in the use of passives to report generalizations occurs in scientific writing. Of the four standard sections of a scientific report — Introduction, Methods, Results, and Discussion — many reports contain a relatively high proportion of passive structures in the two middle sections. This pattern makes sense when we realize that the Method and Results sections should be replicable: the reader should be able to duplicate the author's experiment. Using passive voice helps to convey the impression that the results should hold, regardless of who the agent is. Passage (7), from a Methods section, illustrates this use of passive.

(7)   Each subject was asked to identify and produce a word pair contrasting in final position (e.g., *pat-pad*). Eight productions of each word were recorded in a randomized order, using a Sony D6C cassette recorder. The recorded productions were transcribed by three speech-language pathologists.

In contrast, the Introduction and Discussion sections may have relatively more active structures, since the writer uses these sections to argue for a particular interpretation of the results.

Use a consistent voice to maintain cohesion. As you revise the sentences in a paragraph, you may be able to achieve a smoother flow of information if you keep a single thematic role in the same syntactic position (for example, if you keep the agent of each sentence in subject position). As an illustration, consider passage (8a).

(8a) INCONSISTENT: The project will consist of three steps. First, we will analyze Belcor's financial statements. Second, financial ratios will be compared with the industry norm. Third, the net tax effects of each alternative will be evaluated. Fourth, we will design an implementation plan for the best alternative.

The four steps described here all have the same agent. However, the agent occupies subject position in the first and fourth steps, but is omitted altogether from the second and third steps. The flow of the passage could be improved by using either passive or active voice consistently, as in revisions (8b) and (8c).

(8b) CONSISTENT (ACTIVE): First, we will analyze Belcor's financial statements. Second, we will compare financial ratios with the industry norm. Third, we will evaluate the net tax effects of each alternative. Fourth, we will design an implementation plan for the best alternative.

(8c) CONSISTENT (PASSIVE): First, Belcor's financial statements will be analyzed. Second, financial ratios will be compared with the industry norm. Third, the net tax effects of each alternative will be evaluated. Fourth, an implementation plan for the best alternative will be designed.

For other ways to achieve cohesion, see Chapter 5, "Revising for Cohesion."

Applications

1. Rewrite each sentence to emphasize the patient and de-emphasize the agent.

   EXAMPLE: We have located numerous reviews of each product.
   REVISION: Numerous reviews of each product have been located.

   a. Once we have completed a preliminary analysis, we will submit a formal proposal.

   b. One should obtain information about a company through a study of leading indicators.

   c. Extensive use of computers throughout American life has enhanced the growth of the computer software industry.

   d. Researchers use the AGID test to determine the proportion of potassium in an animal's circulatory system.

2. Rewrite each sentence to emphasize the agent and de-emphasize the patient. Supply an agent if the sentence does not have one.

   a. Research on the current costs associated with both methods will be included in my final report.

   b. The printer should be disconnected from the power source before the cartridge is removed.

   c.   Three samples were tested and ranked by a trained lab technician.

   d.   Those programs have not been defined as retirement accounts by the IRS.

3.   Rewrite each passage so that a single thematic role (agent or patient) appears in a single syntactic position (subject or object).

> EXAMPLE: The manufacturer of each software package was requested to send a demonstration program of its product. Our team then reviewed and analyzed these. The strengths, weaknesses, and cost of each package were evaluated.
>
> REVISION: A demonstration disk for each software package was requested from the manufacturer. Each package was then reviewed and analyzed by our team. The package's strengths, weaknesses, and cost were evaluated.

> a. The lease-buy analysis will include several types of research. We will gather current wheat prices from market reports. Land prices will be obtained from realtors and we will also examine government publications for them. Equipment and machinery prices will be gathered from local implement dealers. We will determine variable input needs such as seed and fertilizer requirements by using yield data from seed companies and the Tennessee Crop Reporting Service. We will obtain prices for these inputs from local suppliers. We should complete this phase of the project by the end of four weeks.

b.   Methanol presents some obvious shortcomings for aircraft. Rubber fuel-system parts are attacked by the alcohol content. Less energy is generated by methanol than by avgas. And methanol tends to suspend impurities in the fuel.

c.   The participants were randomly seated at tables, with either three or four at a table. The test booklets were administered to them, making sure each participant at the tables had a different test booklet from the others. After the experimenter read the introduction to them, the participants were asked to read the text page. They were given 10 minutes to carefully read and study the text. The experimenter then instructed the participants to take the multiple-choice test. They were asked to work alone, to not look back at the text page, and to answer every question to the best of their ability. The participants completed the test and survey questions and turned in their test booklets. A debriefing statement was given to them when they turned the test in.

d.   I conducted most of my research by using journal articles as a way of defining what community policing is and also to get some ideas about programs that could be implemented at the local level. I also conducted some primary research in the city of Plainview to find out if its new community policing tactics have had any noticeable effects on the community. I did this by randomly interviewing 15 citizens of Plainview to see how they feel about the Plainview Police Department and whether or not they have noticed any change in officer behavior since the department's switch to community policing about one year ago. I also interviewed three officers from the Plainview Police Department to see how they feel about community policing, and I also asked them if they use the strategies when they are out on patrol.

4.   Revise Appendix 2 ("Chris Applegate's Job Application Letter") for more appropriate use of active and passive voice.

5.   Revise Appendix 7 ("Clinical Guidelines for Electrosurgery") for more appropriate use of active and passive voice.

Chapter 11. Revising Word Choice

Readers expect professional writing to be accurate, clear, and consistent with professional and standard usage. Therefore, the goal of revising word choice is to identify words that are inaccurate, unclear or misleading, unprofessional, or inconsistent with standard usage. Identifying and revising word choice problems can be difficult because they have a number of different causes. This chapter discusses six common sources of word choice problems. You may also want to review Chapter 8, "Revising for Conciseness," and Chapter 12, "Revising to Improve Tone," both of which deal further with revision at the word level.

Connotation

CONNOTATION refers to an association that many readers attach to a word, but that is not part of its actual definition (or DENOTATION). For example, consider the nouns *agreement* and *deal*. Both words have similar denotations, since both can be used to refer to a contractual agreement between two parties. Therefore, it might seem that both words could be used in a context like *This _____ should benefit our companies.* However, the word *deal* has connotations that make it unsuitable for many professional situations: it may suggest a dishonest arrangement or one in which one party has an advantage over the other. (Note that Donald Trump's book *The Art of the Deal* would have quite a different flavor if he had called it *The Art of the Agreement*.) Problems with connotation often result from using a thesaurus that merely lists synonyms without explaining more subtle differences in their usage. If you are going to use a thesaurus, consider investing in a good one that offers this additional information.

Choosing the appropriate connotation can also depend on audience and culture. For example, consider the difference in tone conveyed by the terms *junkie, drug addict*, and *victim of substance abuse*. Although all three terms might be used to describe the same person, they would not convey the same attitude or degree of objectivity on the writer's part. In the same way, different cultures do not always attach the same connotations to the same words. For example, in British English, the nouns *plan* and *scheme* might be used interchangeably. In American English, though, *scheme* denotes devious behavior.

Jargon

JARGON is the technical vocabulary used in a particular field. To decide whether the use of jargon is appropriate in a given text, you need to consider your audience's degree of familiarity with the specialized vocabulary. When reader and writer share the same technical vocabulary, jargon can serve a useful purpose, acting essentially as "shorthand" for a complex set of concepts. For example, if both you and your reader are familiar with desktop publishing, you can use the jargon term *kerning* with no ill effects. On the other hand, if all or part of your audience consists of nonspecialists, it's best to define a jargon term the first time you use it (e.g., ***Kerning*** *involves adjusting the spacing between pairs of letters to create a more pleasing visual effect*).

Slang and idiomatic terms

The use of slang or idiomatic terms may result in an overly informal style. Sometimes these phrases are appropriate for informal, oral communication, but sound inappropriate in formal writing. For instance, a sales manager might casually tell a salesperson that *Our sales are off the chart this month*, but this same utterance would probably not be formal (or specific) enough to use in a written report. The writer would be better off using a more standard, specific statement such as *Our sales increased 28% this month*.

Subcategorization

SUBCATEGORIZATION problems usually arise from using an object with a verb that does not require one (or, conversely, from omitting an object when a verb requires one). For example, the verb *warn* requires a direct object, a principle violated by the sentence *The consultants warned about the hostile takeover bid*. (Compare *The consultants warned management about the hostile takeover bid*.)

Selectional restrictions

A SELECTIONAL RESTRICTION on a verb means that only certain types of nouns can serve as its subject or object. For example, the direct object of the verb *educate* must be a human noun. This restriction is violated in the sentence *Physical therapists must educate a variety of disabilities*. The sentence can be revised in several ways to introduce a human direct object: for example *Physical therapists must educate a variety of disabled people* or *Physical therapists must educate people with a variety of disabilities*.

Confusing word pairs

A final source of word choice problems arises from the numerous confusing word pairs in English. These include words that sound similar but that have different spellings and usages (e.g., *their/there/they're*; *cite/sight/site*; *comprised/composed*).

Applications

1. Identify and revise the word choice problem in each of the following sentences.

    EXAMPLE — FROM A RESEARCH SURVEY: The author overlooks the fact that many of the occupations held by graduates do not necessarily require a college education.
    ANALYSIS: Violates selectional restriction; an *occupation* cannot be *held*. Revision: The author overlooks the fact that many of the positions held by graduates . . .

    a.   FROM A PROPOSAL: Under the new system, there will be less erosion of the pavement and therefore less potholes and rough spots to tear up the cars as they pass.

    b.   FROM AN EXECUTIVE SUMMARY: Operation Simplification eliminated a lot of bureaucratic garbage that was creating unnecessary paperwork.

    c.   FROM AN ABSTRACT: Although controlled fires are usually designed to serve a single purpose, they frequently benefit several.

    d.   FROM A REPORT: The severity of injury and the amount of wages the employee will lose due to injury can vary compensation.

    e.   FROM A REPORT: It may take some time to gain the same product notoriety in the foreign market that we have in the U.S.

    f.   FROM A PROPOSAL: As we discussed last week, there will be no fee for this report. However, I do have personal expenses that will need to be refurbished.

    g.   FROM A LETTER OF APPLICATION: In addition to the necessary educational knowledge, I also possess certain job qualities. While working at Ace Electric, I learned much about the electrical trade and feel this experience would give me a feel for the job you're offering. This job gave me insight on how to use authority and how to interact with different people. I was also often asked to assist others I worked with use Lotus and WordPerfect.

h.   FROM A REPORT: The company's principle customers are middle-aged and older men in the Eastern U.S.

i.   FROM A BOOK REVIEW: On some topics, the author provides little or no examples, and on others he drones on and on redundantly.

j.   FROM A BOOK REVIEW: The author sights the population increase in Early Modern Europe as one cause of standardized production.

k.   FROM AN ARTICLE: Of these nine items, the majority of the respondents participated in the environmentally conscience behavior for only three.

2.   Identify the jargon terms in each passage. Explain the circumstances under which the use of jargon would be appropriate or inappropriate in each passage, noting hypothetical audiences that would be comfortable or uncomfortable with the jargon terms. Then re-write each passage to eliminate jargon. Some of the technical terms are defined below; you should be able to locate the others in a good dictionary.

dactyl (noun): finger or toe
imprest fund (noun): petty cash
journalize (verb): to make an entry in a journal

a.   The reproductive cycle of the cat is dependent on 12 hours of light for normal cyclicity. Coitus may take place many times, with different males, during estrus.

b.   The mouse showed signs of dactyl paresis after surgery.

c.   The liquidity of assets, including the balance of the imprest fund, should be journalized.

d.   The Seller and Purchaser hereby acknowledge and confirm that this contract states the entire agreement between the parties. The Seller and Purchaser both acknowledge that if any matter as enumerated herein has been of concern to them, they have sought and obtained independent advice.

3.   For each term below, identify at least one other term that conveys the same meaning but carries a different connotation. What differences in attitude are conveyed by the various terms for the same item?

a.   *alcoholic*

b.   *hyperactive child*

c.   *economically disadvantaged*

d.   *set-aside program*

e.   *downsize*

f.   *senior citizen*

3.   Revise Appendix 5 ("A Definition of Aphasia") for more effective word choice.

4.   Revise Appendix 6 ("Memo Assessing an Employee's Writing"), Parts A and B, for more effective word choice.

5.   Revise Appendix 10 ("Report on 'Improving Employee Training at Becker Foods'") for more effective word choice.

Chapter 12. Revising to Improve Tone

TONE refers to the impression that a text conveys about the writer's attitude toward the reader and the topic being discussed. For example, a text may have a personal or an impersonal tone, depending on the amount of distance it conveys between reader and writer. Revising for the right tone is especially important in business correspondence, since this type of writing must establish a positive working relationship with the reader.

This chapter looks at four variables that affect the tone of a text: FORMALITY, READER-ORIENTATION, DIRECTNESS, and PRESUPPOSITION. (Chapter 11 deals with some other elements of word choice that may also affect tone.)

Formality

Like the other elements of tone, the level of formality can be visualized as a continuum.

HIGHLY INFORMAL ← BALANCED → HIGHLY FORMAL

For most routine correspondence, a balanced level of formality is appropriate. This level is businesslike but establishes a more personal relationship between reader and writer by coming closer to the sound of face-to-face conversation. At the same time, it's useful to know how to achieve a more formal tone for those occasions when it's appropriate. For example, you may want to vary your tone depending on how you and your reader differ on several dimensions.

YOU CAN BE LESS FORMAL WITH:	YOU MAY WANT TO BE MORE FORMAL WITH:
• a reader you have previously met or corresponded with • a reader who works for your company • a reader who is about your age (or younger) • a peer or subordinate	• a reader you have not previously met or corresponded with • a customer, client, or reader who works for another company • a reader who is significantly older than you • a reader who is above you in the corporate hierarchy

The rest of this section reviews five ways to vary your level of formality.

Contractions. One way to achieve a balanced style is to use occasional contractions, joining a subject pronoun with an auxiliary verb (e.g., *We'll* from *We will*) or an auxiliary verb with *not* (e.g., *We won't* from *We will not*). In business correspondence, contracting at every opportunity will often lead to an overly informal tone. At the other extreme, avoiding contractions altogether may lead to an overly formal tone that sounds stiff and distant.

Personal pronouns. For a balanced level of formality in business correspondence, use personal pronouns occasionally. The balanced form below (1a) uses three personal pronouns:

you, *your*, and *us*. In contrast, the highly formal version (1b) refers to neither reader or writer.

(1a) BALANCED: You will receive your check from us next week.
(1b) HIGHLY FORMAL: A check will be issued next week.

While personal pronouns are usually appropriate in routine correspondence, under some circumstances you should avoid making the reader the subject of a sentence. These special cases are treated later in this chapter in the sections on reader-orientation and directness.

Plain language. Choosing a shorter or more familiar word over a longer or less familiar word generally creates not only a balanced tone but also an easier-to-read style. As you revise, look for multisyllabic, Latinate words that you can replace with their Anglo-Saxon equivalents. Some examples are given here.

HIGHLY FORMAL: *require, desire, demonstrate/illustrate, be cognizant of,*
                              prioritize, obtain/acquire, utilize
BALANCED: *need, want, show, know, rank, get, use*

Of course, using nothing but plain words may lead to a monotonous, "See Spot run" style. And certainly you may need to introduce (and define) less familiar vocabulary when writing about specialized subjects. For everyday letters and memos, though, plain language usually works.

Placement of prepositions. Some structures (e.g., certain questions and relative clauses) are formed by moving the object of a preposition to the front of a clause. Fronting the preposition as well creates a more formal tone: for example, *On which committees do you serve?* Leaving the preposition in clause-final position creates a less formal tone: for example, *Which committees do you serve on?*

Use of *who/whom*. Using *whom* in objective case creates a more formal tone: *Whom did you hire?* is more formal than *Who did you hire?* If the pronoun is the object of a preposition, use *whom* with a fronted preposition and *who* with a clause-final preposition.

(2a) MORE FORMAL: With whom did you meet?
(2b) LESS FORMAL: Who did you meet with?

In an effort to be highly formal, some writers mistakenly use *whom* in subject position. However, this creates an ungrammatical structure: *Whom will be attending the meeting?*

Reader-Orientation

In most writing directed to a specific audience, you will want to place the appropriate focus on the reader. (This focus is often referred to as *YOU*-PERSPECTIVE or *YOU*-ORIENTATION, where the reader is the *you*.) Reader-orientation can be achieved by revising both sentence structure and the overall content of the text.

Make the reader the subject of positive messages. If the reader will react positively to your letter or memo, try making a reader-oriented pronoun (*you* or *your*) the subject of the sentence that contains the positive news. This strategy is used in (3).

(3a) DRAFT: We will be issuing your bonus check next month.
(3b) REVISED: You will receive your bonus check next month.

Avoid making the reader the subject of negative messages. If the reader will react negatively to your letter or memo, avoid *you* or *your* as the subject of the sentence that contains the negative message. This strategy is used in (4b), which further de-emphasizes the negative message by placing it in a dependent clause.

(4a) DRAFT: Your grant proposal was not selected for funding.
(4b) REVISED: Although we were unable to fund your grant proposal, we invite you to apply again next year.

Make the reader the subject if the reader's action is required. When issuing a policy statement or preparing instructions, use subject position to specify the agent of the action you are describing. Avoid agentless passive structures, which can lead to uncertainty because they fail to assign responsibility. (For review in identifying agentless passives, see Chapter 10, "Revising Active and Passive Voice.") Examples (5a) and 5b) show an agentless passive and its revision.

(5a) DRAFT: Time sheets for student workers should be sent to Payroll every Friday.
(5b) REVISED: The Department Head should send time sheets for student workers to Payroll every Friday.

Address the reader's needs, concerns, and interests. In addition to revising sentence structure, you can increase reader-orientation by including content that reflects the reader's needs, concerns, and interests. For example, (6a), from the introduction to a report, is writer-oriented instead of reader-oriented.

(6a) DRAFT: My report discusses a major strategic decision facing the Allen Corporation: whether to open more retail stores. I will discuss the advantages and disadvantages of opening more retail stores. My recommendation is that we open three new stores in the East.

Instead of focusing on the writer's role in preparing the report, the passage should explain how the report will help the reader make a decision. The revision in (6b) shifts the emphasis to the reader.

(6b) REVISED: The Allen Corporation faces a major strategic decision: whether to open more retail stores. This report evaluates the advantages and disadvantages of opening more retail stores, based on management's goal of regaining a 7% market share from Belker Limited. Based on the evidence presented in this report, the Allen Corporation should open three new stores in the East.

Explain negative messages to the reader. When denying a request for information, credit, an adjustment, or a favor, the challenge is to maintain the reader's goodwill by using a courteous and reasonable tone. One way to maintain goodwill is to explain the refusal to the reader. The five general strategies below can be adapted to explain many negative messages. For example, suppose you are the sales manager at a nursery and have to deny a customer's order for 10 apple trees. Among the reasons you might offer are the following:

EXISTENCE: All or part of the item requested doesn't exist.
    (*We no longer carry apple trees.*)
AGENCY: Someone else is responsible for the item requested.
    (*We sell only on a wholesale basis, but Anderson Nurseries can fill your order.*)
TIMING: Now is not the best time to do the action requested.
    (*Your apple trees are more likely to flourish if you purchase and plant them two months from now.*)
READER BENEFIT: The action requested would not benefit the reader.
    (*Apple trees do not normally thrive in this climate; we recommend that you purchase fig trees instead.*)
ABILITY: The writer is unable to perform the action requested.
    (*Because of unprecedented demand, we have only two apple trees left in stock. However, we offer pear trees at the same price.*)

Directness

The appropriate level of directness is especially important if you are making a request of the reader. The most direct way to state a request is as a plain imperative: for example, *Submit your report by Friday.* Clearly, though, this would create the wrong tone for many readers. This section discusses when and how to revise for a more indirect tone.

Assessing the appropriate level of directness. As a rule, you need a more indirect tone if the "weight" of your request is great. Three variables come into play here.

- How much does the request impose on the reader?
- How unfamiliar are you and the reader with each other?
- How much higher than you in the professional hierarchy is the reader?

If the answer to all three questions is "Not much," then the request is relatively "light," and you can be more direct. At the other end of the scale, if the answer to all three questions is "A great deal," then the request is relatively "heavy" and you need a more indirect tone.

For example, suppose you are placing a routine order with a vendor that you've been doing business with for some time. This request does not impose on the reader, whose job is to fill orders. Nor are you addressing an unfamiliar reader or one whose power exceeds yours. In this situation, a relatively direct tone is entirely appropriate: for instance, *Please bill the following items to our account*.

Now suppose you are asking a special favor of a reader you do not know personally and who outranks you in the professional hierarchy — for example, asking a renowned specialist in your field to answer a complex inquiry. Here, you would want to be more indirect: for instance, *I would greatly appreciate it if you could answer the following questions*.

Achieving indirectness. You can make a request more indirect by adding elements to a plain imperative or by using a non-imperative structure. Below are seven strategies for increasing indirectness, each used to vary the plain imperative *Submit your report by Friday*. The italics in each example highlight the sentence part that implements the strategy.

STRATEGY	EXAMPLE
• Use a deference marker	*Please* submit your report by Friday.
• Phrase your request as a question	*Could you* submit your report by Friday?
• Add an uncertainty marker	*If possible*, I need your report by Friday.
• Generalize your request	*All reports* are due by Friday.
• Nominalize your request	The deadline for report *submission* is Friday.
• Incur a debt	*I would appreciate* having your report by Friday.
• Apologize for the request	*Sorry for the short notice*, but I need your report by Friday.

You may have noticed that the indirect forms are all longer than the direct form (*Submit your report by Friday*). This is no coincidence, since in most languages there is a trade-off between conciseness and abruptness: it usually takes more words to state a request indirectly.

Of course, too many indirectness strategies in one sentence can sound like groveling: *I'm terribly sorry to have to ask you this, but I wonder if you could possibly submit your report by Friday*. Used judiciously, though, these strategies can soften an overly direct tone.

Presupposition

A PRESUPPOSITION TRIGGER is a word that conveys either a truth judgment or a value judgment. As such, a presupposition trigger conveys an assumption that the reader may not agree with and, therefore, can affect tone.

Wh-words. A *wh*-word (*who, what, when, where, who, how*) presupposes the truth or certainty of a proposition in the same sentence. For example, *When will the merger occur?* presupposes that "The merger will occur." In contrast, the *yes/no*-question *Will the merger occur?* does not convey the same presupposition. Neither does the conditional structure *If the merger occurs, we may be open to antitrust charges*.

Factive verbs. A factive verb is one that presupposes the existence or truth of the clause or noun phrase that follows it. For example, *Brown demonstrated the superiority of his method* conveys the writer's belief that "Brown's method is superior," since the factive verb *demonstrate* presupposes the truth of its object. On the other hand, *Brown claims that his method is superior* neither affirms nor denies the writer's belief that "Brown's method is superior," since *claim* is non-factive. Below is a partial list of factive and non-factive verbs.

> FACTIVE: demonstrate, note, acknowledge, prove, show, grasp, make clear, be aware, take into account, bear in mind, regret, resent
> NON-FACTIVE: assert, claim, suppose, allege, assume, charge, maintain, believe, think, conclude, conjecture, fancy, figure

Implicative verbs. An implicative verb presupposes a judgment about some action described in the sentence. For example, compare (7a) (non-implicative) with (7b) and (7c) (implicative).

(7a) You did not enclose a check.
(7b) You forgot to enclose a check.
(7c) You failed to enclose a check.

Unlike *did not*, the implicative verbs *forget* and *fail* convey a judgment about the fact that the reader did not enclose a check. *Forget* implies that the reader had an obligation, but unintentionally did not meet it. *Fail* also implies that the reader had an obligation, but without the mitigating factor of unintentionality. Therefore, (7c) attaches more blame to the reader than (7c). For this reason, *fail* may create an undesirable tone.   Below is a partial list of other implicative verbs and their presuppositions.

> *neglect:* unmet obligation (e.g., *You neglected to enclose your check*).
> *remember:* met obligation (e.g., *Jones remembered to fax the report*).
> *avoid:* negative act (e.g., *We should avoid unnecessary risks*).
> *refrain:* negative act (e.g., *Please refrain from smoking at your desk*).
> *manage:* intentional, difficult act (e.g., *I managed to finish the report*).
> *happen:* unintentional, easy act (e.g., *I happened to hear your conversation*).

Applications

1. Revise each of the following sentences to an appropriate level of formality for a report to an executive audience.

> EXAMPLE: State University offers a really great MBA program.
> REVISION: State University offers a highly respected MBA program.

   a. Epstein's article shows how asbestos cases are going through the roof.

   b. This idea probably would not work well because the company would need incredible increases in unit sales due to the low profit margin on the product.

   c. The Associated General Contractors are hopping mad about the state's plan to award $450 million in design-build contracts over the next eight years.

2. Analyze the different presuppositions conveyed by the sentences in each sets.

> EXAMPLE: The results of the ratio analysis make clear that WMX is a stable company.
> ANALYSIS: The factive verb *make clear* presupposes that "WMX is a stable company."

   ia. Has your order been lost?
   ib. When was your order lost?
   ic. You claim that your order has been lost.
   id. We regret that your order has been lost.
   ie. If your order has been lost, please contact us.

   iia. I happened to find the information that you requested about accounting software.
   iib. I managed to find the information that you requested about accounting software.
   iic. I neglected to find the information that you requested about accounting software.

3.   Revise each sentence for the appropriate reader-orientation, depending on whether its message is negative, positive, or one that directs the reader to do something.

EXAMPLE: We will send your order by the date you specified.
REVISION: You will receive your order by the date you specified. (positive)

a.   You will not receive the increase in credit you requested.

b.   You will not receive your printer for another 6 weeks.

c.   The purchase price of your stereo will be refunded.

d.   We want all expense sheets before June 30.

e.   When comparing HVT measurements, one must be extremely careful that all parameters are exactly the same as those used before.

4.   As the assistant manager of personnel at ChemAg, Inc., you have received a résumé and a request for an interview from a chemical engineer named Karl Langley. Compose refusals using two different strategies for explaining negative messages.

5.   Revise each of the following requests in three different ways to increase indirectness.

EXAMPLE — FROM AN EMPLOYEE TO A SUPERVISOR: Give me a vacation.
REVISIONS: Could I have a vacation? (Question)
       All employees are supposed to take their vacations by July 1. (Generalize)
       I'd really appreciate a vacation. (Incur a debt)

a.   FROM AN EMPLOYEE HANDBOOK: Wear your ID card while you are in the building.

b.   FROM THE PRESIDENT OF THE CHAMBER OF COMMERCE TO A LOCAL BUSINESS EXECUTIVE: Speak at our next meeting.

c.   FROM A LETTER OF APPLICATION FOR A JOB: Call me at (555) 482-7652 after 5:00 p.m. to arrange an interview.

6.   Revise the following memo for more appropriate tone. Assume it is being sent to residents of an apartment building by the manager of that building.

REMINDER TO ALL RESIDENTS: SMOKING POLICY

Do not smoke in any of the public areas within the building. This includes the entries, stairwells, corridors, and laundry rooms. Advise your guests of our building policy and tell them to refrain also. The policy is in effect for the health and safety of all residents and will be strictly enforced.

7.   The following passage is from a set of guidelines about how to design or select control devices for instrument panels. Revise the passage so that it uses a more appropriate tone. Focus on level of formality and reader-orientation.

In the design or selection of control devices, it is important to consider whether the movement and location of the control device and the element to be controlled are compatible. The physiological and anatomical efficiency with which the operator can use the control should also be considered. In relation to compatibility, controls should be oriented to fit normal habit-patterned reflexes. The location of controls which are most often used must be placed between elbow and shoulder height for physiological efficiency. In addition, control dimensions should take into consideration the normal hand-grasp limitations.

8.   The following letter was received by a reader who was attempting to recover the cost of damages to her car after it hit a loose piece of cement on a state highway. Rewrite the letter to improve its tone, focusing on level of formality and reader-orientation.

> Your small claim against the Department of Transportation and Development has been reviewed pursuant to Acts (1977) No. 596, R.S. 13:5141 et seq. as amended. With regard to your claim, it is believed that there is no liability on the part of the Department of Transportation and Development.
>
> State law provides that we must have notice of a defective condition prior to an accident, and also have a reasonable time to make repairs when prior notice is present. If either or both of these factors is missing, there is no liability on the part of the State.
>
> Investigation of your claim has indicated that the factors which produce liability are not present. In view of this, your claim for settlement is being denied under the Small Claim law cited above. This letter will serve as formal notice of denial.

9.   Revise Appendix 2 ("Chris Applegate's Job Application Letter") to improve its tone.

10. Revise Appendix 3 ("Brian Carter's Job Application Letter") to improve its tone.

11. Revise Appendix 6 ("Memo Assessing an Employee's Writing"), Parts A and B, to improve their tone.

Chapter 13. Editing Punctuation

This chapter focuses on the uses of punctuation that most often present problems in professional writing. The items discussed are the period, comma, semicolon, colon, parentheses, brackets, hyphen, and apostrophe, and the use of numbers.

Period

The most common use for a period is, of course, to signal the end of a sentence. In addition, though, the period serves a number of other uses in professional writing. Most of these involve abbreviations of words or phrases.

Courtesy titles. Use a period to abbreviate a courtesy title such as *Mr.*, *Mrs.*, *Ms.*, *Dr.*, and *Prof.* when it is followed by a name: for example *Ms. Ellen Barlow*. However, spell out words like *doctor* or *professor* when they are not used before a name, as in (1).

(1)   Dr. Stirling, who was recently named Turner Professor of Medicine, was selected by a committee of seven doctors.

Degrees. Use periods to abbreviate degrees such as *B.A.*, *B.S.*, *M.A.*, *M.B.A.*, *M.S.*, *M.D.*, *J.D.*, and *Ph.D.* (When a degree is used as a person's title, it is separated from the name by a comma). Examples (2) and (3) illustrate these conventions.

(2)   Pat Murray, M.D.
(3)   He recently received a Ph.D. in French.

Some fields vary in this usage. For example, you may see *Certified Public Accountant* abbreviated as either *C.P.A.* or *CPA*. Generally, the form with periods reflects more conservative usage.

Other abbreviations and acronyms. A phrase is often abbreviated to the first letters of the major words in the phrase: for example, *NBC* for *National Broadcasting Company*. An ACRONYM is an abbreviation that forms a pronounceable word: for example, *NASA* from *National Aeronautics and Space Administration*. The tendency in punctuating both abbreviations and acronyms is toward omitting the periods between the abbreviated items.

Units of measurement are also abbreviated frequently in technical writing. In general, you can omit the period unless the abbreviation could be confused with another common word. The examples in (4) illustrate these usages.

(4)   5 in.   19 kg   8-1/2 lb
      389 mph   4 mHz

In an address, use a period after any abbreviated form, except that for the state. The state is abbreviated as two capital letters with no punctuation, as in (5).

(5) Ms. Ellen Formstead
1918 E. 31st St., Apt. 901
Fresno, CA 90028

The abbreviations *e.g. (for example)* and *i.e. (in other words)* have a period after each letter, and they are usually used in parentheses. A comma follows either abbreviation.

(6) It may be more efficient to use a macro instead of a function key (e.g., Alt-i instead of Ctrl-F8-2-4).

Comma

The basic function of a comma is to separate items within a sentence. Within this basic function, there are four main uses for the comma.

To separate items in a series. Place a comma after each item in a series of two or more. We advise using a comma before the last item in a series, since confusing structures may result if it is omitted. For example, the comma after *Engineering* in (7) makes it immediately clear that Engineering and Math are separate buildings, not one.

(7) Renovations are planned for the Humanities, Physics, Engineering, and Math buildings.

If one of the items in a series contains internal commas, as in (8), use a semicolon instead of a comma to separate the items.

(8) The participants included Dr. Chris Adams, Head of Engineering; Dr. Robert Brigham, Vice-Chancellor; and Dr. Caroline Selwyn, Dean of Liberal Arts.

To separate introductory material from a main clause. Common types of introductory material are prepositional phrases (a preposition followed by a noun phrase), dependent clauses (which begin with a subordinating conjunction such as *because* or *although*), and participial phrases (which often begin with an *-ing* verb form). Use a comma to signal the end of the introductory material and the beginning of the main clause, as in (9)-(10).

(9) In the last chapter, DeWitt presents his conclusions.
(10) Because she had read about the company, she was prepared for the interview.

To separate independent clauses joined by a coordinating conjunction. When two independent clauses are joined by a coordinating conjunction (*and, but, or, nor, so*), use a comma after the first clause; see (11)-(12).

(11) The results are inconclusive, but they suggest several areas for further research.
(12) You can request an advance for travel, or you can submit receipts after the trip.

To separate a conjunctive adverb from a clause. Conjunctive adverbs such as *however*, *therefore*, *nevertheless*, *consequently*, *as a result*, and *furthermore* typically occur either at the beginning of a clause or between elements in a verb phrase. When a conjunctive adverb begins the clause, as in (13), use a comma after it. When it occurs within the clause, as in (14), use a comma before and after it.

(13) Three subjects received incomplete instructions. Consequently, they were unable to complete the task.

(14) The committee sets policy. The director, however, implements it.

Semicolon

The main use of the semicolon is to join two independent clauses, as in (15)-(16).

(15) The committee sets policy; the director implements it.

(16) Over half of our employees chose Delmar Dental; the others chose either Northern Dental or Telcor.

If the second clause contains a conjunctive adverb (e.g., *however*, *therefore*, *nevertheless*, *consequently*, *as a result*), it goes after the semicolon, as in (17)-(18).

(17) The committee sets policy; the director, however, implements it.

(18) We have experienced growth during the last three quarters; therefore, we should be able to absorb this loss.

A semicolon also separates items in a series when one or more of the items contains internal punctuation, as in (19).

(19) The participants included Dr. Chris Adams, Head of Engineering; Dr. Robert Brigham, Vice-Chancellor; and Dr. Caroline Selwyn, Dean of Liberal Arts.

Colon

The main function of a colon is to link general or introductory material to specific or explanatory material. Some common uses of the colon are illustrated below.

To introduce a list. Use a colon between an independent clause and a list that follows it in the same sentence. Example (20a) illustrates this use.

(20a) The following shipping methods are available: Federal Express, Greyhound Package Express, United Parcel Service, and U.S. Postal Service.

However, if the list serves as the complement or direct object of the verb, no colon is needed unless the list is in a vertical format. For example, no colon is needed in (20b); the colon is optional in (20c).

(20b)   The available shipping methods are Federal Express, Greyhound Package Express, United Parcel Service, and U.S. Postal Service.

(20c)   The shipping methods we can choose from are:
    1.   Federal Express
    2.   Greyhound Package Express
    3.   United Parcel Service
    4.   U.S. Postal Service

To introduce a definition or explanation. A colon can separate an independent clause from a definition or explanation that follows it, as in (21).

(21)   The new advertising campaign was designed with one primary market segment in mind: single women from 25 to 35 years old.

To introduce an extended quotation. Use a colon to link your comments about an author to an extended quotation from the author, as in (22). (Note that an independent clause introduces the quotation.)

(22)   Chomsky defines generative grammar as a model of abstract linguistic knowledge rather than a model of speech production: "When we say that a sentence has a certain derivation with respect to a particular generative grammar, we say nothing about how a speaker or hearer might proceed, in some practical or efficient way, to construct such a derivation."

To introduce a subtitle. In a bibliographical entry, use a colon to separate the main title of a work from its subtitle, as shown in (23).

(23)   Newmeyer, F. J. (1980). *Linguistic Theory in America: The First Quarter-Century of Transformational Generative Grammar*. New York: Academic.

You can also use this format on the title page of a report or article, as in (24).

(24)   Landfill Needs in Myerville:
                      A Plan for the Next Decade

To set off parts of letters and memos. Colons follow the headings on a memo.

(25) Date: December 13, 1998
     To: All Department Heads
     From: Gale Storm, Director of Personnel
     Re: New Search Committee Guidelines

A colon also follows the attention line on a letter, as in (26).

(26) Attn: Order Department
     Shelray Electronics
     14 Industrial Park Road
     Baton Rouge, LA 70802

The salutation on all but the most informal business correspondence is also followed by a colon.

(27) Dear Ms. Sandusky:

A colon is also used in the enclosure or copy line at the bottom of correspondence.

(28) encl: Check #1480
(29) c: Karen Bell, Director of Marketing

Parentheses

Parentheses are used to enclose words, phrases, clauses, or sentences that are subordinate to the sentence or paragraph in which they occur. When using parentheses within a sentence, no punctuation goes immediately before the left parenthesis. Punctuation may follow the right parenthesis, as in (30a).

(30a) Prices for three carriers are given below (prices for Premier Courier have been excluded since their location is not within an three-mile radius).

When using parentheses around an entire sentence, put the period inside the parentheses, as in (30b).

(30b) Prices for three carriers are given below. (Prices for Premier Courier have been excluded since their location is not within a three-mile radius.)

Brackets

Brackets have two primary uses. One is to replace parentheses in a text that already occurs in parentheses, as in (31).

(31) Select F1 to view the picture in more detail. (You may have to scroll the document [i.e., press Page Down] to bring the picture into view.)

The second common use of brackets is to enclose comments or explanatory material within a passage that you are quoting directly from another source. Using brackets helps to distinguish your comments from those of the writer you are quoting. This strategy is used in (32), where the bracketed material was added by the writer quoting the manual.

(32) According to the manual, "It is inadvisable to mix serif typefaces [i.e., those with `feet'] with sans serif faces [those without `feet']."

Occasionally, you may have to quote a passage that contains a typographical or factual error. If so, use the word [*sic*] in brackets after the error to tell the reader that the error was in the original material. For example, in the original passage quoted in (33), *data* was incorrectly printed as *date*.

(33) Smith states that "Overvoltage surge can cause scrambled date [*sic*] transmission."

Dash

The dash (formed by typing two hyphens) has uses that overlap with those of the colon, parentheses, and comma. It can be used introduce or set apart explanatory, defining, or parenthetical material, as in (34)-(35).

(34) Taste tests on meat that has been irradiated--a procedure that kills E. coli and other bacteria--were conducted using a double-blind procedure.
(35) Declines in the frog population have been attributed to several causes--chemical pollutants, ultraviolet radiation, and disease-causing algae.

Dashes have a somewhat less formal flavor than alternative forms of punctuation, so it's best to use them sparing in professional writing.

Hyphen

Modifiers and nouns in professional writing often occur in compound forms, with two or more words functioning as an adjective, adverb, or noun. Such compounds enable more detailed descriptions and can allow you to combine terms in new ways — an option especially important in technical fields, where you may need to construct terms for new concepts. The following principles explain how to hyphenate some common types of compounds.

Compound adjectives. Hyphenate the parts of a compound adjective when it precedes a noun. However, don't hyphenate between the compound adjective and the noun.

(36) well-organized proposal   twentieth-century invention
     30-second alarm   2-inch pipe
     computer-generated image   entry-level position

Do not hyphenate adverb-adjective compounds if the adverb ends with *-ly*.

(37) recently remodeled facilities   newly developed theory
     poorly written proposal   overly complicated form
     heavily traveled route   badly needed innovation

Compound nouns. In general, do not hyphenate compounds functioning as nouns.

(38) A number of species have become extinct during the twentieth century.
(39) Test the alarm every 30 seconds.

Numbers. Hyphenate prose (i.e., spelled-out) names of numbers from 21 to 99 and prose fractions. When using numerals, hyphenate whole and fractional numbers that appear on the same line.

(40) Forty-three units   one-fifth of the faculty
     8-1/2 minutes   9-2/5 inches

You can also use a hyphen to resolve ambiguous groupings of three or more words, as in (41).

(41a) 8-1/2 minute intervals [i.e., an unspecified number of intervals, each lasting for 8-1/2 minutes]
(41b) 8 1/2-minute intervals [i.e., 8 intervals, each lasting for 1/2 minute]

Apostrophe

In *it's*. Use the apostrophe in *it's* only if you can substitute the words *it is*. Examples (42)-(45) illustrate this principle.

(42) It's clear that this plan is better than that one.
(43) The sample is large, but it's not representative.
(44) This model has one major weakness: its rudder is unstable.
(45) This model has one major weakness: its unstable rudder.

In other contractions. Use an apostrophe to contract a noun-verb or verb-negative sequence, as in (46)-(47).

(46) You'll receive a free ticket after you've flown 20,000 miles.
(47) I haven't seen the results yet.

In other possessive forms. Use an apostrophe to signal the possessive form of a noun.

(48a)   The company's earnings-per-share ratio is below industry average.
(49a)   About 60 percent of a probation officer's time is spent in preparing paperwork.

When the possessive form is based on a plural, put the apostrophe at the end of the form.

(48b)   The companies' earnings-per-share ratios are below industry average.
(49b)   Paperwork accounts for 60 percent of most probation officers' work loads.

Note, however, that no apostrophe is used in a possessive pronoun.

(50) Our proposal was more convincing than theirs.
(51) The best presentation was yours, not his or hers.
(52) Its advantages are clear.

Numbers

Numbers and ordinals (e.g., *first, second*) appear frequently in business and technical writing, since such writing often discusses precise quantities. You should be aware of two types of conventions when using numbers and ordinals: how to choose between prose numbers and numerals, and how to use punctuation within numerals. The following guidelines cover some common situations. As you will see, most of the guidelines recommend using numerals (rather than prose numbers).

Numbers below and above 10. Spell out numbers and ordinals below 10; use numerals for 10 and above. However, if a passage mixes both types, use numerals throughout.

(53) Of the three reports, the third was clearer than the first or the second.
(54) We need to order 2 adding machines, 4 boxes of pens, and 18 calendar refills.
(55) Please review the records for the 8th, 12th, and 22nd subjects.

Numbers that begin a sentence. Spell out numbers that begin a sentence (or revise the sentence so that it does not begin with a number).

(56) Forty-seven of the samples were defective.
(57) So far, 1242 students have registered for College Composition. (Revised from "One thousand-two hundred-forty-two students . . . .")

Adjacent numbers. In a series of two adjacent numbers, spell out either the first number or the shorter one. Use a combination of words and numerals for very large numbers.

(58) twelve 18-month calendar refills   2.2 million
      1701 First Street $276 trillion

Units of measurement. Use numerals with units of measurement, percentages, sums of money, fractions, and decimals, even if the number is below 10. (A contract may contain both prose and numerical forms.)

(59) 2 percent   0.912 oz   5000 Hz   4 cm
      2-1/2%   8 miles $9.42   3-year-old

Use a zero before decimal fractions of less than one.

(60) 0.918   0.018   0.008

Parts of a text. Use numerals to designate parts of a text: e.g., *Section 2, figure 3, page 11, line 4, Chapter 5.*

Times and dates. Use numerals for times and dates: e.g., *The contract expires at 12:00 A.M. on June 15, 1992.*

Applications

Identify and edit any punctuation problems in the following passages. Some passages have more than one problem.

> EXAMPLE: JDM Realtors is the intended user of this report, however, other companies may also find this report informative.
> EDITED: JDM Realtors is the intended user of this report; however, other companies may also find this report informative.

1. Murray's book provides a lot of useful information including: performance graphs, helpful hints, company listings by performance, and general rules for successful investing. The book contains ratings of over four-hundred companies with detailed information on the most important aspects of company performance.

2. The reactor consists of one foot of 6 inch diameter pipe that branches out to 21-eight foot segments of four inch diameter pipe.

3. CAREER OBJECTIVE: Desire entry level position with opportunity to contribute in hazardous waste, process control or design groups. Long term goal is to manage one of these groups.

4. The most common keyboard layout is known as QWERTY, named for the first six characters on the top alphabetic row of keys, the other layout is called Dvorak, named for the developer August Dvorak.

5. Scott and Willits (1994) used ten items (three consumer focused items and 6 politically focused items.)

6. If a new firm gets into trouble, the venture capitalist may have to replace managers, (or even founding entrepreneurs) to get the company back on the road to success.

7.   The new store to store link allows the pharmacist to fill tourists and out of towners prescriptions in a matter of minutes.

8.   With plant expansion a manufacturer may be able to use a tax exempt bond. A disadvantage to the issuance of the bonds is the nonrenewal status of the debt, if the interest rates go down the company cannot simply refinance the debt.

9.   Well written users manuals help a first time user to learn the features of a package. They also help the experienced user to look up seldom used features when needed.

10.   An area of concern in the software industry, is the expected shortage of skilled computer programmers and system's analysts.

11.   Although two thirds of all physical therapists continue to be employed by hospitals; private clinics and nursing homes also employ, and draw upon the skills of physical therapists.

12.   Investors can earn 6% on the new bonds, however, they could earn 8% on the old bonds.

13.   Because the turbocharger forces more fuel and air into the combustion chamber more power is produced when it is burned.

14.   The number of elderly is already large and with the baby boomer population moving towards retirement there will be a large increase in the number of people who will be faced with decisions about care for the elderly.

15. As competition has increased and firms have become mass producers of small products some industries have moved away to the less complicated job order system.

16. Individual members, gang cliques, or entire gang organizations traffic in drugs, commit shootings, assaults, robbery, extortion, and other felonies, and terrorize neighborhoods.

17. The observation and notes were mainly from the boys mother who relayed information about her children to the case worker.

18. In 1995 Minneapolis reached a record of 97 homicides, police attributed 80% of them to gang related violence. (Nelson 1997:1A).

19. This case shows how an effective early childhood program can change a child's behaviors, however without proper follow through the changes may not be permanent.

20. It would help the citizens perception of the police department if the police could respond faster to public calls for service. The police department response time could be lowered significantly if people wouldn't wait an average of 17 minutes to call the police for help after a crime has been committed. If the public had a better and more open relationship with the police department maybe they would call for help faster which would result in more criminals being apprehended. Also if the departments budget could be expanded to include about three more cars on patrol each day then the police could be more visible to the people which would make them feel safer and more protected.

21. Edit Appendix 3 ("Brian Carter's Job Application Letter") for punctuation.

22. Edit Appendix 10 ("Report on 'Improving Employee Training at Becker Foods'") for punctuation.

Chapter 14. Editing Sentence Fragments

Incomplete sentences occur frequently in informal conversations (e.g., *Where did she eat? At the deli*). Incomplete sentences usually do not create a problem in conversation because each speaker provides a syntactic context for the other's utterance. For example, the complete sentence *She ate at the deli* can be reconstructed from context. In contrast, professional writing must be clear and complete because the writer isn't on hand to answer questions for the reader. In addition, standard written English requires more formality than does casual conversation. For these reasons, your final draft must be free of SENTENCE FRAGMENTS.

This chapter explains how to identify and edit common types of sentence fragments. Most fragments can be edited in two ways: by joining the fragment to a sentence (usually the one right before it), or by adding elements to make the fragment a complete sentence.

Fragments consisting of phrases

One type of fragment (italicized in each example below) results from punctuating a noun phrase, verb phrase, or prepositional phrase as a sentence. One way to edit a fragment of this type is to re-punctuate it as part of another sentence, as in (1)-(3).

(1a) FRAGMENT: We should encourage employee participation in cost control. *For example, an award for the best money-saving idea.*

(1b) EDITED: We should encourage employee participation in cost control: for example, an award for the best money-saving idea.

(2a) FRAGMENT: Many new companies have entered the industry. *Making it more difficult to gain market share.*

(2b) EDITED: Many new companies have entered the industry, making it more difficult to gain market share.

(3a) FRAGMENT: This material has many uses. *For instance, to repair tissue.*

(3b) EDITED: This material has many uses — for instance, to repair tissue.

Alternatively, you can add elements to make the phrase a complete sentence.

(1a) FRAGMENT: We should encourage employee participation in cost control. *For example, an award for the best money-saving idea.*
ANALYSIS: This fragment needs a verb phrase for the noun phrase, *an award*.

(1c) EDITED: We should encourage employee participation in cost control. For example, we could give an award for the best money-saving idea.

(2a) FRAGMENT: Many new companies have entered the industry. *Making it more difficult to gain market share.*
ANALYSIS: This fragment needs a subject. Also, a present participle (*-ing*) verb can never function as the only verb in a sentence.

(2c) EDITED: Many new companies have entered the industry. These newcomers are making it more difficult to gain market share.

(3a) FRAGMENT: This material has many uses. *For instance, to repair tissue.*
ANALYSIS: This fragment needs a subject. Also, an infinitive phrase (*to*-verb) can never function as the only verb in a sentence.

(3c) EDITED: This material has many uses. For example, surgeons can use it to repair tissue.

Fragments consisting of dependent clauses

A DEPENDENT CLAUSE contains a noun phrase and a verb form, but also begins with a subordinating conjunction that prevents it from functioning as a sentence (that is, as an INDEPENDENT CLAUSE). Because a dependent clause contains a noun phrase and a verb phrase, it may be difficult to identify as a fragment. About the only solution to this problem is learning to recognize some common subordinating conjunctions. The following list illustrates some subordinating conjunctions categorized in terms of meaning.

CAUSE-EFFECT: *because, if, since*
TIMING OR SEQUENCE: *after, as long as, as soon as, before, once, until, when*
COMPARISON: *as, just as*
CONTRAST: *although, despite the fact that, even though, in spite of the fact that, though, while, whereas*

Of these, *because, although, when,* and *if* are among the most commonly used.
The simplest way to correct a dependent clause fragment is to join it to the beginning or the end of the independent clause that usually comes before it, as in (4).

(4a) FRAGMENT: Student retention has risen. *Although new student enrollment has declined.*
(4b) EDITED: Student retention has risen, although new student enrollment has declined.

(4c) FRAGMENT: Student retention has risen. *Because of increased advising efforts.*
(4d) EDITED: Because of increased advising efforts, student retention has risen.

A word of advice: don't stop using subordinating conjunctions in order to avoid fragments. As discussed in Chapter 6, "Revising to Build Transitions," these conjunctions give the reader useful cues about the relationship between ideas in different clauses. The main principle to remember is to link a dependent clause with an independent clause.

Fragments consisting of present participle clauses

A PRESENT PARTICIPLE CLAUSE contains a noun phrase immediately followed by an *-ing* verb form. You can edit a present participle clause either by changing the verb so that the *-ing* form does not stand alone, or by combining the clause with the preceding sentence. Examples (5) and (6) show examples of this type of fragment, along with edited versions.

(5a) FRAGMENT: The mail-order industry has several advantages. *One of those being the ability to target catalogs at specific demographic groups.*

(5b) EDITED: The mail-order industry has several advantages. One of those is the ability to target catalogs at specific demographic groups.

(6a) FRAGMENT: Each program makes different use of the input devices. *Some uses being more logical than others.*

(6b) EDITED: Each program makes different use of the input devices, some uses being more logical than others.

Fragments consisting of relative clauses

A RELATIVE CLAUSE fragment begins with a RELATIVE PRONOUN (*who, whom, which,* or *that*), followed by a noun or a verb form. To edit, either replace the relative pronoun with another pronoun or a noun phrase, as in (7b) or join the relative clause to the preceding sentence, as in (8b).

(7a) FRAGMENT: The company has hired a number of employees over the past six months. *Who may be laid off after the merger.*

(7b) EDITED: The company has hired a number of employees over the past six months. They may be laid off after the merger.

(8a) FRAGMENT: The meeting will be simulcast from the Twin Cities. *Which will allow viewers in Duluth to watch.*

(8b) EDITED: The meeting will be simulcast from the Twin Cities, which will allow viewers in Duluth to watch.

Fragments consisting of indirect questions

An INDIRECT QUESTION is a clause that begins with a *wh*-word (*who, what, when, where, how*) and is immediately followed by a noun rather than a verb phrase. An indirect question fragment can be edited in several ways.

- Join the indirect question to the preceding sentence (often with a colon or dash).
- Make the indirect question the object of a verb or of a preposition.
- Change the indirect question to a DIRECT QUESTION by placing the first auxiliary verb after the *wh*-item.

Example (9) illustrates an indirect question and two ways to edit it.

(9a) FRAGMENT: The task force should recommend recycling procedures. *In particular, what steps we can take to reduce paper waste.*

(9b) EDITED: The task force should recommend recycling procedures. In particular, we need to know what steps we can take to reduce paper waste.

(9c) EDITED: The task force should recommend recycling procedures. In particular, what steps can we take to reduce paper waste?

Applications

Identify and correct the fragment in each passage.

> EXAMPLE: The company has two branches. One in West Knoxville and the other downtown.
> EDITED: The company has two branches: one in West Knoxville and the other downtown.

1. Also known as stockholders' equity, book value is the difference between the company's assets and its liabilities. In other words, what the shareholders own after the company's debts have been paid.

2. There have been several breakthroughs recently in cash register software. A major one being that the receipt printer can now take the place of the old rebate system.

3. The project will require an additional $48,600 in equipment expenses. Over the amount proposed in the original budget.

4. For teachers to develop new lesson plans in the sciences will require increased support from school administrators and school boards. For example, providing additional professional development classes for teachers.

5. A first course of action to consider is to try to maintain profitability, by putting together a political lobby group and sending them to lobby the British government. Trying to get them to change one or both of their proposed actions.

6. The company should move into a more specialized product area, yet remain with their original catalogs, too. For example, special catalogs for children's equipment and clothing, or a winter sports catalog.

7. Under the proposed system, we would have to pay employees for only a total of six sick days a year. Resulting in a yearly savings of $43,968.

8.   As we discussed earlier, the report will consist of two major sections. The first of these will look at expansion into chain stores and factory outlets. Second, an in-depth look at sales potential where the competition is currently prevailing.

9.   One option available to employees is a non-contributory five percent pension plan. Which basically means that the company will contribute five percent of your $30,000 salary.

10.   According to one source, it is better for a corporation to be a multinational company, rather than strictly a domestic company. The reason being that a strictly domestic company is tied solely to the economy of one country, but a multinational firm is tied to the economies of many different nations.

11.   A consultant with expertise in the retail field will have to be hired. Who knows what type of promotional campaigns to run and where certain items would best be located in the store. The consultant would have to be employed a minimum of one week and perhaps as long as a month or more. Although the cost associated with employing a consultant will be offset in the long run by having a better program.

12.   My research will familiarize the investor with several different aspects of franchising. How much capital is needed for specific franchises, the most often used profit-sharing methods, and the policies that must be followed to keep the franchise.

13.   Parents are being held responsible for contributing to the academic development of their children. While teachers are becoming accountable for instructing a more diverse classroom.

14.   Edit Appendix 6 ("Memo Assessing an Employee's Writing"), Part B, for fragments.

15.   Edit Appendix 10 ("Report on 'Improving Employee Training at Becker Foods'") for fragments.

Chapter 15. Editing Modifiers

A MODIFIER is a word, phrase, or clause that gives more information about another item in the sentence. Typically, modifiers function as adjectives (giving more information about a noun) or as adverbs (giving more information about an adjective or verb). For example, in the sentence *A newly convened committee of four faculty met on Thursday*, *newly* modifies *convened*; *newly convened* and *of four faculty* modify *committee*; and *on Thursday* modifies *met*. This chapter reviews ways to edit four common modifier problems in technical and business writing.

Avoiding ambiguous modifiers

Ambiguity arises if the reader is unclear about which item the modifier goes with. Ambiguity is especially a problem with prepositional phrases and relative clauses. Since these items have some flexibility in where they are placed in a sentence, they may be too far away from the element you intend them to modify. For example, consider (1a).

(1a) AMBIGUOUS: We would like to talk to you about starting a new job next month.

The prepositional phrase *next week* is functioning as an adverb. The problem is that the sentence contains two possible verbs that the adverb could modify: *talk* and *starting*. To avoid confusion, the writer needs to revise so that only one interpretation is possible.

(1b) EDITED: We would like to talk to you next month about starting a new job.
       We would like to talk to you about a new job that would start next month.

Ambiguity also arises in (2a).

(2a) AMBIGUOUS: Gifts may not include tobacco or perfume costing more than $5.00.

The ambiguity here comes from the adjectival phrase *costing more than $5.00*. Does it apply to both *tobacco* and *perfume*? Or does it apply to *perfume* alone? To ensure the first interpretation, the writer could edit as in (2b).

(2b) EDITED: A gift of tobacco or perfume may not cost more than $5.00.

To restrict the modifier to *perfume* alone, the writer could edit differently, as in (2c).

(2c) EDITED: Gifts may not include perfume costing more than $5.00 or tobacco.

The word *only* may also create modification problems. Because it can function as either an adjective or adverb, *only* needs to be placed carefully, in front of the item it modifies. For example, consider the effects of moving *only* from one slot to another in the following sentence.

___(1)___ that dealer ___(2)___ sells ___(3)___ pre-owned Cadillacs.

Placing *only* in slot (1) implies "No other dealer sells pre-owned Cadillacs." Placing *only* in slot (2) implies "That dealer sells pre-owned Cadillacs but doesn't perform other services such as repairing them." Placing *only* in slot (3) implies "That dealer sells a single type of car: pre-owned Cadillacs." Try constructing a similar variety of readings for the sentence below.

___(1)___ two subjects ___(2)___ developed a rash ___(3)___ after ___(4)___ an hour of exposure to the substance.

Avoiding "dangling" modifiers

Typically, a DANGLING MODIFIER results from a combination of two conditions:

- the sentence begins with an introductory phrase containing a present participle (-*ing*) verb form with no noun phrase in front of it, and
- the subject of the main clause is not a logical agent for the -*ing* verb.

For example:

(3a) DANGLING: While inspecting the plant, several safety violations were noted.

Upon reading the introductory phrase *While inspecting the plant*, the reader looks for an agent — the "doer of the action" *inspecting*. Since there is no agentive noun phrase in the introductory material, the reader tries to interpret the subject of the main clause (*several safety violations*) as the agent. Obviously, though, this won't work, since *inspecting the plant* requires a human subject but *several safety violations* is nonhuman. (See Chapter 11, "Revising Word Choice," for more on such "selectional restriction" violations.) Thus, *While inspecting the plant* is a dangling modifier — it is unanchored to any noun phrase in the sentence.

There are two simple ways to edit a dangling modifier. One is to introduce the agent of the -*ing* verb into the introductory phrase. This approach is used in (3b).

(3b) EDITED: While the engineers were inspecting the plant, several safety violations were noted.

The other approach is to make the agent of the -*ing* verb the subject of the main clause. This approach is used in (3c).

(3c) EDITED: While inspecting the plant, the engineers noted several safety violations.

Note, by the way, that (3d) still contains a modification problem:

(3d) While inspecting the plant, several safety violations were noted by the engineers.

Although this version introduces an agent (*the engineers*) into the main clause, the agent is still not in subject position.

Watch too for potential ambiguity when using an introductory phrase. For example:

(4a) UNCLEAR: By filling out the enclosed forms, we will be able to make the changes you want in your policy.

Who's supposed to fill out the form here? Actually, the reader is. Structurally, though, the writer (*we*) is the agent of the clause that the introductory phrase modifies. In order to avoid this confusion, the writer could edit as in (4b) or (4c).

(4b) EDITED: After you fill out the enclosed form, we will be able to make the changes you want in your policy.

(4c) EDITED: By filling out the enclosed form, you can specify the changes you want in your policy.

Avoiding "stacked" modifiers

STACKED MODIFIERS usually occur when an adjective-noun sequence functions as an adjective to modify another noun, as in (5a).

(5a) STACKED: The small business economic outlook does not look promising.

The sequence *small business economic outlook* creates a problem because the reader has a hard time deciding whether each noun is the subject of the sentence or simply a modifier of the subject. You can edit stacked modifiers by placing some of them after the noun, as in (5b).

(5b) EDITED: The economic outlook for small businesses does not look promising.

Avoiding wordy or redundant modifiers

Sometimes modifiers, especially prepositional phrases and relative clauses, can be condensed or omitted without sacrificing information. Consider (6a).

(6a) WORDY: Two pipes, six feet in length, feed the coolant into the reactor.

The phrase *in length* can be either condensed or omitted: for example, *Two 6-foot-long pipes feed the coolant into the reactor.* Similarly, the information in the relative clause in (7a) can be stated more concisely.

(7a) WORDY: The report which was written by the Accounting Department identifies three problems in our inventory system.

(7b) EDITED: The Accounting Department's report identifies three problems in our inventory system.

Applications

Identify and correct the modification problem in each passage. In some cases, you may have to supply information (e.g., a subject noun phrase).

> EXAMPLE: By having more products in stock, our shipping accuracy should improve to the desired 97.5%.
>
> EDITED: By having more products in stock, we should be able to improve our shipping accuracy to the desired 97.5%.

1.   By using the Universal Price Code, delays due to misticketing and no tickets are almost totally eliminated.

2.   In order to create a balanced reading program, whole language materials must be supplemented with phonics materials.

3.   By self-insuring one's company, doors are opened to lawsuits and even possible bankruptcy.

4.   A prescribed burning prepares the land for future forests. By reducing weak competition and exposing the mineral soil, the land becomes very fertile for growth.

5.   Though presented clearly, Monroe provides little data to support his claims.

6.   After grading each package on each criterion, a clear winner became evident.

7.   Hopefully after reading my research paper, there is a better understanding of the involvement of children in gangs.

8. After determining company stability, the stock itself will be analyzed to determine if the selling price at the current time seems reasonable.

9. By interviewing administrators in the Education Department, this report presents information about upcoming curriculum changes.

10. By working as a waitress at Chi Chi's, it has enabled me to support myself through college.

11. Providing practice in writing and recognizing new vocabulary, first- and second-grade children can keep their own dictionary of new words they have learned.

12. Several locations were found to place the proposed plug-flow reactor after reviewing the plant blueprints. While physically inspecting the plant site, information about the operating conditions for the proposed reactor and distillation columns was gathered. By assuming a pipe diameter of 4 inches, the total length of the reactor was calculated to be 164.1 feet, which breaks down to 21 eight-foot segments.

13. Under this program, a team of civilian gang outreach workers maintains regular contact with up to 200 known gang members, providing them with crisis counseling, job and school referrals, and other alternatives.

Chapter 16. Editing for Subject-Verb Agreement

A subject and verb AGREE if the verb form reflects the number of the subject (singular or plural). The number of the subject changes the form of the verb *to be* in both present and past tense, as shown here.

	SINGULAR		PLURAL	
	PRESENT	PAST	PRESENT	PAST
1ST PERSON	I am	I was	We are	We were
2ND PERSON	You are	You were	You are	You were
3RD PERSON	He is	She was	They are	They were

In all other verbs, only a present-tense verb with a third-person singular subject changes form. Examples (1)-(4) illustrate this change.

THIRD-PERSON SINGULAR	THIRD-PERSON PLURAL
(1a) He *has* no relevant experience.	(1b) They *have* no relevant experience.
(2a) The study *supports* our conclusion.	(2b) The studies *support* our conclusion.
(3a) She *does* a good job.	(3b) They *do* a good job.
(4a) This number *goes* in that column.	(4b) These numbers *go* in that column.

As you can see, the third-person, singular form of a present-tense verb ends in *-s* (e.g., *is, has, supports, does, goes*). (Note that nouns use the opposite pattern: plural forms end in *-s*).

While subject-verb agreement is relatively straightforward in simple sentences like the examples above, it can be somewhat trickier in more complex constructions. This chapter reviews some constructions that often create agreement problems.

A subject containing a prepositional phrase

Sometimes a subject contains a head noun (i.e., the noun that the verb must agree with) followed by other material such as a prepositional phrase. Examples (5)-(6) show how to edit this type of structure for subject-verb agreement.

(5a) FAULTY: The essays in this newly published *book supports* our claim.
(5b) EDITED: The *essays* in this newly published book *support* our claim.

(6a FAULTY: His new article on animal sleeping *habits appear* in *Zookeeper*.
(6b) EDITED: His new *article* on animal sleeping habits *appears* in *Zookeeper*.

A subject containing a relative clause

The head noun of the subject is sometimes followed by a relative clause. Examples (7)-(8) show how to edit this type of structure for subject-verb agreement.

(7a) FAULTY: The essays that will appear in her next *book supports* our claim.

(7b) EDITED: The *essays* that will appear in her next book *support* our claim.

(8a) FAULTY: An advisory committee, which includes the *deans and department heads, meet* every term with the vice-chancellor.

(8b) EDITED: An advisory *committee*, which includes the deans and department heads, *meets* every term with the vice-chancellor.

A subject containing coordinated items

If a subject contains two or more noun phrases joined by *and*, use a plural verb form.

(9) *The manager, the account executive, and the client have* been invited to meet.

If two noun phrases in a subject are coordinated by *either-or* or *neither-nor*, use a verb that agrees with the second noun phrase.

(10a) SINGULAR + SINGULAR: Neither the manager nor the *client was* informed.

(10b) SINGULAR + PLURAL: Neither the manager nor the *clients were* informed.

(10c) PLURAL + PLURAL: Neither the managers nor the *clients were* informed.

(10d) PLURAL + SINGULAR: Neither the managers nor the *client was* informed.

Other elements in subject position

A gerund. A GERUND is an *-ing* verb form functioning as a noun. Use a singular verb with this type of subject, unless it ends in *-s*.

(11) *Meeting* with clients *is* the account executive's responsibility.

(12) Several *meetings* with the client *are* scheduled.

A title. Use a singular verb if the subject is a title.

(13) *Teaching Tips contains* an article about designing homework assignments.

A unit of measurement. If the subject is a unit of measurement equated with a singular complement, use a singular verb form.

(14) Given his credit record, *$10,000 is* a reasonable line of credit.

(15) *Five years is* the normal probationary period for untenured faculty.

(16) For most people, *80 decibels is* an uncomfortable noise level.

The word *there*. If the subject is *there*, use a verb that agrees with the noun phrase that follows the verb.

(17) There *are problems* with this proposal.

Applications

Identify and correct the agreement problem in each passage.

> EXAMPLE: Vast amounts of research has been expended in these areas.
> EDITED: Vast amounts of research have been expended in these areas.

1.   The difference between these two methods is the way that factory overhead (all other costs besides direct material and labor) are accounted for.

2.   DeRoyal has doubled in size in the last two years, and with this growth the demands on the current computer system is very great.

3.   As the amount of ozone depletion caused by the chemicals address in the Montreal Protocol are lowered, the global percentage of ozone destruction caused by SRMs will increase.

4.   My electives in upper-division finance includes Investment Analysis, and my accounting classes have given me a good financial academic background.

5.   The final report, including charts, appendixes, and indexes, are scheduled to be completed by December 1.

6.   The core of the system — notification and record keeping — were developed over a nine-month period.

7.   When Quality Circles work, there is a direct benefit to management in terms of cost and time. But there is also indirect benefits to the employees.

8.   The qualifications for drafting assistant, as listed in your advertisement, includes several areas in which I have experience.

9.   Either surgery or physical therapy are prescribed for knee injuries.

10.   Under this program, a team of civilian gang outreach workers maintain regular contact with up to 200 known gang members.

11.   Eating disorders among teenagers and young women is a growing problem in this society.

12.   Educating ourselves about women's programming needs, feminist theory, and ethics are a must to know and be aware of.

13.   Edit the executive summary of Appendix 10 ("Report on 'Improving Employee Training at Becker Foods'") for subject-verb agreement.

Chapter 17. Editing Pronoun Reference

A pronoun, by definition, is a word used in place of another noun phrase (called the ANTECEDENT or REFERENT). For example, in the sentence *A writer should avoid jargon if it will confuse the reader*, *it* refers to the antecedent *jargon*. Pronouns in English take a variety of forms, depending on their person (first, second, or third) and case (nominative, objective, or possessive). Third-person singular pronouns also vary according to gender (masculine, feminine, or neuter). The chart below lists the personal pronouns in English.

	NOMINATIVE		OBJECTIVE		POSSESSIVE	
	SINGULAR	PLURAL	SINGULAR	PLURAL	SINGULAR	PLURAL
1ST PERSON	I	we	me	us	my	our
2ND PERSON	you	you	you	you	your	your
3RD PERSON	he/she/it	they	him/her/it	them	his/her/its	their

English also contains the demonstrative pronouns *this, that, these*, and *those*, and the relative pronouns *who(m), that*, and *which*.

Pronoun reference enables the reader to identify an antecedent for each pronoun. As you edit, you should be aware of three goals related to pronoun use: clear reference, agreement between the pronoun and its antecedent, and gender-neutral pronoun use.

Editing for clear reference

Since English has a small number of pronouns and a large number of nouns, very often a pronoun may have more than one possible antecedent. In other words, the pronoun's reference may be ambiguous or unclear. This lack of clarity can cause problems for the reader, who needs to be able to identify a single referent for each pronoun.

In order to give each pronoun a single referent, you may need to rearrange or add sentence parts. In (1a), for example, *he* has two possible referents, as revealed by the two edited versions.

(1a) UNCLEAR: *The client* sued *the broker* after *he* invested unwisely.
(1b) EDITED: After *the client* invested unwisely, *he* sued the broker.
After *the broker* invested unwisely, the client sued *him*.

Likewise, the pronoun *it* has three possible referents in (2).

(2) UNCLEAR: If you try to connect *the computer* to *the printer* with *the wrong driver*, you may damage *it*.

Try constructing three edited versions, each with a different referent for *it*.

Demonstrative or relative pronouns may also have unclear antecedents. Since these pronouns can refer to entire clauses or sentences, you may need to go back a sentence or two to locate the antecedent. For example, the referent for *this* in (3) is vague.

(3) UNCLEAR: Phillips had the art department design three new layouts. He then spent an hour discussing them with the client. *This* probably kept us from losing the account.

The referent for *this* may be specific; for example, *This meeting*. Or it may be more general; for example, *This extra effort*. In either case, the referent needs to be clarified.

Editing for pronoun-antecedent agreement

Use a singular pronoun to refer to a singular antecedent and a plural pronoun to refer to a plural antecedent. While this principle is straightforward, problems in number agreement often arise if the antecedent is an indefinite pronoun (e.g., *each, someone, everyone*) or a collective pronoun (e.g., *company, committee*), or if the antecedent and pronoun are in different clauses.

Indefinite pronouns. In conservative usage, indefinite pronouns are generally treated as singular: for example, *Each account executive has his or her own sales number*; *Everyone has his or her own sales number*. There is some evidence that this usage is changing; in less conservative usage, you may see *their* used with an indefinite antecedent: *Each account executive has their own sales number*; *Everyone has their own sales number*. However, you should be aware that more conservative readers may object to this usage. (The section below on "Using Gender-Neutral Pronouns" offers other alternatives to the *his or her* pattern.)

Collective nouns. Collective nouns, at least in American English, are generally treated as grammatically singular, even though they refer to an entity comprised of a number of individuals. Edit to maintain a consistent pronoun number when referring to a collective noun.

(4a) INCONSISTENT: The committee was asked to design a more cost-effective plan. *It* met once a week for two months. *They* also interviewed six plant managers.
(4b) CONSISTENT: The committee was asked to design a more cost-effective plan. *It* met once a week for two months. *It* also interviewed six plant managers.

Antecedent in a different clause. Be careful not to shift from singular to plural (or vice versa) if the antecedent and pronoun are in different clauses.

(5a) INCONSISTENT: A cochlear implant is not a cure for deafness and can only provide *a child* with the sensation of sound; *they* must use other visual means of gaining information.

(5b) CONSISTENT: A cochlear implant is not a cure for deafness and can only provide *children* with the sensation of sound; *they* must use other visual means of gaining information.

Editing for gender-neutral pronouns

Traditionally, the singular masculine pronouns (*he, him,* and *his*) were used to refer to indefinite pronouns and generic antecedents, as in (6)-(8).

(6a) *Each student* should bring *his* exam book to the final.
(7a) *A manager* needs to meet regularly with *his* employees.
(8a) *The average American* saves 5.7% of *his* salary.

Understandably, however, many readers object to the implied gender bias in such usage (especially, if both male and female students, managers, and Americans are being referred to!). As a result, many professional organizations have adopted guidelines for non-sexist language. In order to avoid the appearance of gender-biased language, you should be aware of some strategies for avoiding the singular masculine pronoun when the antecedent may include female referents.

Replace the pronoun. One strategy is to replace *his* with *a, the*, or no pronoun at all, as illustrated here.

(6b) *Each student* should bring *an* exam book to the final.
(7b) *A manager* needs to meet regularly with employees.

Use *his or her*. Another option is to use *his or her* in place of *his*. For example:

(6c) *Each student* should bring *his or her* exam book to the final.
(7c) *A manager* needs to meet regularly with *his or her* employees.
(8b) *The average American* saves 5.7% of *his or her* salary.

This solution works fine for limited stretches of text (i.e., a sentence or two). However, if you use *his or her* frequently throughout a longer text, it's likely to become distracting. Therefore, you may want to use a different approach in a longer text.

Pluralize the antecedent. For dealing with longer texts, the best strategy is often to put the antecedent in plural form, thus eliminating the need for a singular pronoun. The examples here illustrate this approach.

(6d) *All students* should bring *their* exam books to the final.
(7d) *Managers* need to meet regularly with *their* employees.
(8c) *Americans* typically save 5.7% of *their* salaries.

Note that this solution may create some ambiguity if you pluralize the object of the possessive pronoun. For example, is each student expected to bring one exam book, or more than one? If the context does not clarify the exact meaning, you may want to further edit the sentence to avoid this ambiguity.

Applications

Identify and correct the problem in pronoun reference in each sentence or passage.

> EXAMPLE — FROM A DISCUSSION ON SHOPLIFTING: In most instances, the customer is led to believe that he is responsible for most shrinkage a store experiences. In fact, however, shoplifting is a very small percentage of the total.
>
> EDITED: In most instances, customers are led to believe that they are responsible for most shrinkage a store experiences. In fact, however, shoplifting is a very small percentage of the total. (Original version does not use gender-neutral pronouns.)

1.   FROM A PROPOSAL TO AN INVESTOR: The rapidly changing computer industry makes investment difficult for someone who does not keep their knowledge current. In addition, it is very important for an investor to match the type of stock they are looking at with the type of portfolio they are building.

2.   FROM A DISCUSSION OF GOVERNMENT PROGRAM EVALUATION: Full-scale program evaluation requires an examination of all agency/department operations and activities to determine how well it is achieving goals or objectives.

3.   FROM A REPORT ON ATTENTION DEFICIT DISORDER: Attention Deficit Disorder and Attention Deficit Hyperactivity Disorder are both disorders of the central nervous system. It interferes with a child's school work, and can affect their life at home.

4.   FROM A REPORT ON A COMPANY'S INSURANCE NEEDS: When a company has coverage through an insurance company, they are said to be **indemnifying** themselves. Michigan Insurance Service has covered Martec's insurance needs for five years now, but as most companies faced with financial problems would do, they are looking into various ways to cut back on its current and future costs. These include studying the possibility of being self-insured.

5. FROM A DISCUSSION OF A NOTIFICATION SYSTEM USED FOR A UNIVERSITY EXIT EXAM: Notification cards are created with a threefold purpose. Initially, each card contains the Exit Exam ID assigned during the selection process so that a student can enter it on the Exit Exam answer scan form. Secondly, they alert the Testing Center to how many students to expect. Finally, it indicates if the student took the Entrance Exam as a freshman. If the student took the Entrance Exam as a freshman, a special message is produced on the notification card to insure that their Entrance and Exit Exam scores are compared.

6. FROM AN ARTICLE ON FEAR EXTINCTION: Many cognitive-behavioral therapists use the behavioral technique of extinction. This allows the therapist to show the client their specific fears. This process is repeated over and over until the client becomes extinct to the fear. It allows the client to come to terms with their fears. The therapist trains their client to tolerate their fears and express them in a healthy way.

7. FROM A REPORT ON HOW TO EXPAND MAIL-ORDER SALES: The mail-order industry has some very distinct advantages, one of those being the ability to send a catalog to prospective buyers who fit the demographics of a "typical purchaser." This is an opportunity for our company to tap into a rapidly growing market segment such as the women's apparel market. Women's catalogs could be marketed for the same seasons that the existing catalogs are. Separate Christmas issues, for example, would present more items to a more targeted market. The disadvantage to this is the higher cost of printing and postage. It would need to be determined.

A winter sports catalog would be perfect for MaxWear to go into because of the main regions it sells to. They are very big in the Northern and Mountain regions where they can take advantage of outdoor winter sports. They can market clothing and accessories for downhill, cross-country, and telemark skiing. These should be offered to both men and women and to skiers of all abilities.

Chapter 18. Editing References to Other Sources

One of the most persuasive types of evidence for supporting a claim in professional writing is a reference to other sources of information. Consider examples (1a) and (1b), used to illustrate the use of evidence in Chapter 2:

(1a) ORIGINAL: People who are comfortable with and knowledgeable about computers will more likely be creative and work faster than those who feel restricted by limited computer knowledge.

(1b) REVISED: People who are comfortable with and knowledgeable about computers will more likely be creative and work faster than those who feel restricted by limited computer knowledge. For example, Markowitz (1996) compared the performance of accounting clerks who had undergone an 8-hour training session on a spreadsheet program to that of clerks given a 1-hour training session supplemented by written materials. The clerks who had an 8-hour training session made an average of 52% fewer mistakes and were able to prepare a spreadsheet 3 times faster than the other group.

Version (1b) is more persuasive because another source of information supports the writer's claim.

There are three reasons that you need to provide references to other sources in professional documents.

- First, you are ethically bound to acknowledge the work of those people who actually produced or gathered the information you are using as evidence.
- Second, you may be legally bound by intellectual property laws (e.g., copyright) to acknowledge or even seek permission from the sources of information you are using as evidence.
- Third, you provide a service to your readers who want or need to learn more about a topic by telling them how to locate your sources.

This chapter explains the fundamentals of referencing other sources of information in professional writing. (Sources of more comprehensive discussions are listed in the References at the end of this book.)

Types of sources

When you write professional documents, you will rarely rely only on your own knowledge and experience for all of the evidence you need to support your claims. There are many types of potentially useful sources of information for any project. For instance, suppose you are preparing a feasibility report on implementing a pollution prevention program at the restaurant you manage. The following list is representative of the sources that might be available to you.

- An unpublished report written six months ago about pollution prevention at a restaurant/bar.
- Notes from an interview you did with the manager of another restaurant.
- A pamphlet from a utility company outlining how to save energy costs through proper maintenance and cleaning of refrigeration units.
- A pamphlet from a state government agency about pollution prevention in small business.
- A brochure from a recycling company.
- The URL for a state government-sponsored Web site supporting pollution prevention in the food service industry.
- A report about pollution prevention published in the *Journal of Environmental Health*.
- A book about environmental management strategies for business.

Ways to use other sources

When you use material from other sources in professional documents, you will either quote from that source, as in (2a), or paraphrase from it, as in (2b). You might write either of the examples below in your feasibility report. (Note that both examples follow Toulmin's model of claim-evidence-interpretation as discussed in Chapter 2.)

(2a) QUOTE: Many restaurant owners have reacted positively to pollution prevention programs. As the owner of Guthrie's told me, "deciding to implement waste reduction procedures, and cardboard and glass recycling has been one of the smartest business decisions I've made over the past 22 years" (Terry Stewart, personal interview, July 3, 1998). Thus, pollution prevention in a restaurant/bar makes good business sense.

(2b) PARAPHRASE: Many restaurant owners have reacted positively to pollution prevention programs. The owner of Guthrie's told me that their pollution prevention plan has been very successful (Terry Stewart, personal interview, July 3, 1998). Thus, pollution prevention in a restaurant/bar makes good business sense.

You should quote a source if the original wording will have more impact or cannot be improved upon. Thus, (2a) is probably more effective that (2b). On the other hand, if the original wording is unclear or lacks conciseness, or if you are summarizing the point of a whole section of the source, you will want to paraphrase.

The mechanics of using quotes are illustrated in (2a): quotation marks surround the exact words taken from the other source. If your quote is long (i.e., it occupies three or more lines in your document), use the extended quote format shown in (2c). In either case, always introduce a quotation by linking it with your own words in the same sentence.

(2c) EXTENDED QUOTE. Many restaurant owners have reacted positively to pollution prevention programs. As the owner of Guthrie's told me,

> Deciding to implement waste reduction procedures, and cardboard and glass recycling has been one of the smartest business decisions I've made over the past 22

years. Why, I would just never have believed that we could improve our profit margin so dramatically by taking more control over the waste we're producing. In the last quarter we spent about $500 less on garbage pick-up and earned about that much from selling our cardboard. That's a thousand bucks just by managing cardboard boxes better! (Terry Stewart, personal interview, July 3, 1998)

Thus, pollution prevention in a restaurant/bar makes good business sense.

Note that the extra white space (indentation) takes the place of quotation marks in extended quotes. Note also that there is no punctuation after the parenthetical citation. If the source had page numbers (e.g., a book), you would include the page on which the quote is found, like so: (Stewart, 1998, p. 21).

Documenting other sources

Documenting a source means providing the information needed to locate that source. At a bare minimum, you'll need to provide the following:

- Author (name of individual or corporate author, name of editor, etc.)
- Title (title of report or article, title of book, main heading of Web page, etc.)
- Date (publication date for article, submission date for unpublished report, date of interview, etc.)

You provide these details in two places within your document: in a bibliography at the end of the document and in parenthetical citations within the body of the document. (We will not be discussing footnotes at all, because they are now very rarely used in professional documents.) There are many different documentation systems; the choice of a system depends on your field or the field of your readers. It's also important to use the same documentation system consistently throughout a document. We'll confine our discussion here to two types commonly used in professional writing: the author/year system (represented by APA style) and the number system (represented by IEEE style).

Constructing a bibliography

At the end of the body of your document, you provide a list of the sources you have referred to in your document. Each source appears only once in this list, which is sometimes called "works cited" or "references" as well as "bibliography." There are two ways of organizing such a list of sources: by last name of author (as in APA Style) or by order of mention in the document (as in IEEE Style). The following lists demonstrate each method.

(3a) APA STYLE:
    Alabama Power. (1997). Energy Conservation [Pamphlet]. Birmingham, AL: Author.
    Bansal, P. and Howard, E. (1997). Business and the Natural Environment. Woburn, MA: Butterworth-Heinemann.

California Integrated Waste Management Board. (1998). Food service waste reduction tips and ideas [Online]. Available: http://www.ciwmb.ca.gov/mrt/wpw/wpbiz/fsfood.htm [1998, June 8]. Sacramento, CA: Author.

Craig, M., Howell, M., Kim, J., Nelson, G., and Stewart, T. (1998). <u>Waste Reduction and Recycling Assessment for The Music Hall</u> [Unpublished report]. University of Alabama, Tuscaloosa, AL.

Jones. B. (1997). What is pollution prevention? <u>Journal of Environmental Health, 59</u>, 30-35.

(3b) IEEE STYLE:

[1] B. Jones, "What is pollution prevention?" *Journal of Environmental Health, 59*, pp. 30-35, 1997.

[2] Alabama Power. *Energy Conservation.* [Pamphlet]. Birmingham, AL: Author, 1997.

[3] M. Craig, M. Howell, J. Kim, G. Nelson, and T. Stewart, *Waste Reduction and Recycling Assessment for The Music Hall.* [Unpublished report]. University of Alabama, Tuscaloosa, AL, 1998.

[4] California Integrated Waste Management Board. "Food service waste reduction tips and ideas" [WWW page]. [Cited 1998, June 8]. Sacramento, CA: Author. Available from < http://www.ciwmb.ca.gov/mrt/wpw/wpbiz/fsfood.htm >.

[5] P. Bansal, and E. Howard, *Business and the Natural Environment*. Woburn, MA: Butterworth-Heinemann, 1997.

In APA Style, the sources are arranged alphabetically by the last name of each source's author. In IEEE Style, the sources are numbered and arranged according to the order in which they are mentioned within the body of the document.

The lists above demonstrate the mechanics involved in listing five types of sources in APA and IEEE styles: (a) pamphlets or brochures, (b) books, (c) Web sites, (d) unpublished reports, and (e) published reports appearing in periodicals. The lists also illustrate the mechanics of listing sources with various types of authors: (a) single authors, (b) corporate authors, and (c) multiple authors.

Providing parenthetical citations

There are two ways to use parenthetical citations in professional documents. First, if you are using only one or two other sources of information, you may omit a bibliography and simply put the documentation details in the body as a parenthetical citation.

(4a) WITHOUT BIBLIOGRAPHY: You should check your thermostat for correct calibration. With correct calibration, lowering the thermostat by one degree will reduce energy consumption by 3% in winter; raising the thermostat by one degree will reduce energy consumption by 5% in summer (Alabama Power. [1997]. <u>Energy Conservation</u> [Pamphlet]. Birmingham, AL: Author). This can mean a significant increase in profit margin for a restaurant and bar (Craig, M., Howell, M., Kim, J., Nelson, G., and Stewart, T. [1998]. <u>Waste Reduction and Recycling Assessment for The Music Hall</u> [Unpublished report.] University of Alabama, Tuscaloosa, AL).

Note that a citation is placed within parentheses at the end of *and* inside the sentence containing the information which you found in the source (i.e., the citation goes *before* the period at the end of the sentence). In this way, the placement of the citations in (4a) accomplishes two goals: (a) it makes clear which source contains information about calibrating thermostats because the citation occurs within a specific sentence, and (b) it interferes as little as possible with the reading process because the citation is placed at the end of that sentence.

Second, if you are using more than a couple of sources, you need to use parenthetical citations in addition to your bibliography. These parenthetical citations make clear which source in the bibliography contains the information you provide as evidence within the body of your document.

(4b) WITH BIBLIOGRAPHY (APA STYLE): You should check your thermostat for correct calibration. With correct calibration, lowering the thermostat by one degree will reduce energy consumption by 3% in winter; raising the thermostat by one degree will reduce energy consumption by 5% in summer (Alabama Power, 1997). This can mean a significant increase in profit margin for a restaurant and bar (Craig et al., 1998).

Example (4b), using APA Style, lists only the author and year of publication within the parenthetical citation. Your reader can recover the rest of the details about these two sources from your bibliography. If you use any part of the citation in the sentence itself, then you omit that from the parenthetical citation, as in (4c).

(4c)   You should check your thermostat for correct calibration. According to Alabama Power (1997), with correct calibration, lowering the thermostat by one degree will reduce energy consumption by 3% in winter; raising the thermostat by one degree will reduce energy consumption by 5% in summer. This can mean a significant increase in profit margin for a restaurant and bar (Craig et al., 1998).

Note that the parenthetical citation appears directly after the "author" instead of at the end of the sentence in this case. In addition, only the last name of the primary author is used to document sources with more than two authors. However, *et al.* substitutes for the remaining names in parenthetical citations only; all of the authors should be listed in the bibliography.

IEEE Style dictates that you list only the number of a source in a parenthetical citation, which is enclosed in square brackets, as in (4d).

(4d) WITH BIBLIOGRAPHY (IEEE STYLE): You should check your thermostat for correct calibration. With correct calibration, lowering the thermostat by one degree will reduce energy consumption by 3% in winter; raising the thermostat by one degree will reduce energy consumption by 5% in summer [2]. This can mean a significant increase in profit margin for a restaurant and bar [3].

In IEEE Style, the sources are numbered in their order of mention in the document. If the same source is used to support more than one claim within a document, the parenthetical citation always contains the same number. So, for example, if another sentence were added to the end of (4d), and that sentence used information from the Alabama Power pamphlet, the parenthetical citation in that final sentence would be [2].

With any documentation system, when you decide to write a large section of your document using the information from a single source (this is not necessarily a wise choice!), you can explicitly tell the reader this fact and omit using parenthetical citations in every sentence. The boldface sentence in (5) demonstrates how to accomplish this.

(5)   This section of the report focuses on energy conservation activities which can save a restaurant/bar considerable savings on its utility bill. **The majority of the information included here is contained in a pamphlet published by Alabama Power [2].** First, you should check your thermostat for correct calibration. With correct calibration, lowering the thermostat by one degree will reduce energy consumption by 3% in winter; raising the thermostat by one degree will reduce energy consumption by 5% in summer. Second, you can replace higher wattage light bulbs with lower wattage bulbs if you clean light fixtures at least every six months in order to maximize the amount of light output. This can mean a significant increase in profit margin for a restaurant and bar [3].

Note that the parenthetical citation is used only once, even though there are multiple sentences that report information from the same source. Also note that a parenthetical citation for a different source makes clear when information is used from a different source. If you found the same information in two or more other sources, you would include all citations within one set of parentheses, as below:

(6a) MULTIPLE SOURCES (APA STYLE): Second, you can replace higher wattage light bulbs with lower wattage bulbs if you clean light fixtures at least every six months in order to maximize the amount of light output. This can mean a significant increase in profit margin for a restaurant and bar (Alabama Power, 1997; Craig et al., 1998).

(6b) MULTIPLE SOURCES (IEEE STYLE): Second, you can replace higher wattage light bulbs with lower wattage bulbs if you clean light fixtures at least every six months in order to maximize the amount of light output. This can mean a significant increase in profit margin for a restaurant and bar [2, 3].

If you are citing information found on a specific page or set of pages, list those in the parenthetical citation like so: (Alabama Power, 1997, pp. 3-4; Craig et al., 1998, p. 422) in APA Style and [2:3-4, 3:422] in IEEE Style.

Remember the general principle about referencing other sources of information is that you want to be as clear as possible about where your audience can locate the information used as evidence in your document.

Applications

1.   Devise two other potential placements for the parenthetical citation below. List the advantages and disadvantages of each.

The method outlined by Fox [13] lays out the basic steps involved, and there is no reason to disregard these steps since they seem logical.

2.   Revise the use of parenthetical citations in the following paragraphs in order to interfere as little as possible with the reader's reading process.

a.   At a technical interchange meeting of a Launch Vehicles Working Group, 15-16 September 1994, the conclusions of Chapter 10, "Predicted Rocket and Shuttle Effects on Stratospheric Ozone," from the Scientific Assessment of Ozone Depletion: World Meteorological Organization, 1991 were accepted (World Meteorological Organization, 1991).

b.   Some researchers contend that many forms of graphics do indeed enhance the decision-making process (Horton, 1991, p. 12), while others (Tufte, 1983, p. 107) consider some forms of graphics to be a waste of space.   Still others (Schaubroeck and Muralidhar 1991, p. 127; Jarvenpaa, 1986, p. 3) maintain that empirical evidence neither supports nor rejects either position.

3.   Revise the parenthetical citations in the following paragraphs so that they conform with IEEE or APA Style.

a.   Additionally, Sandman (et al., 1994) states that investigators have identified many factors other than risk magnitudes that seem to influence how the public responds to particular risks.

b.   We patterned this project after a report of a communication audit found in volume 5, issue 2 of The Bulletin, June 1994. The title of the paper was "The Communication Audit as a Class Project" written by Roger N. Conaway. The article discussed the procedures to follow to conduct an audit of communication practices in a work place. We adapted the same procedures to a smaller project.

c. The System/Segment Specification document was prepared to support those who will "Develop hardware, interface, and software requirements specifications." [4, p.47.]

d. Furthermore, Robinson [2] et al., have suggested the possibility of a previously unsuspected heterogeneous process involving alumina particles from SRMs and halocarbon gases that could prove to be catalytic and destroy more ozone than previously suspected.

e. In 1959, H. P. Edmundson and R. E. Wyllys, of the Planning and Research Corporation, compiled a report summarizing several methods of creating automatic abstracts and indexes. It was "understood that a frequency count of the significant words of a document can serve to isolate the special vocabulary used to convey information in any particular realm of discourse." (Edmundson, 1959:34)

4. Revise the following bibliographic entries so that they conform with IEEE or APA Style.

a. Susan M. Katz (1998) Part I — Learning to Write in Organizations: What Newcomers Learn About Writing on the Job in IEEE Transactions on Professional Communication, volume 41, issue 2, June, pp. 107-115.

b. "IMIS Products and Customers." [Unpublished report.] Wright-Patterson Air Force Base, OH: Logistics Research Division, Human Resources Directorate, Armstrong Laboratory, November 1995.

5. Locate those places in Appendix 8 ("Research Proposal on 'Erosion at Sawyer Road Park') where parenthetical citations could be most valuable as evidence for the author's claims.

6. Edit Appendix 11 ("Report on 'Security Methods for Ashley's Clothing Store') so that the parenthetical citations and bibliography conform with either IEEE or APA style.

7. Edit Appendix 12 ("A Discussion of Prenatal Exposure to Cocaine") so that quotations are properly integrated into the text.

Appendix 1. Memo to Residents

(For use with Chapters 4 and 7.)

The following draft is of a memo to residents of an apartment building from the building managers.

MEMO TO ALL RESIDENTS: GARAGE CLEANING

2

    The garages in your building are scheduled to be cleaned starting on Tuesday, August 28

4 and continuing through Saturday morning, September 1. The work will consist of pressure washing all walls, ceilings, and pipes in the garage area. The cleaning is scheduled as

6 follows: Tuesday August 28 stalls 13 through 25 beginning at 9:00 a.m. and finishing at 9:00 p.m. Wednesday, August 29 stalls 26 through 37 beginning at 9:00 a.m. and

8 finishing at 9:00 p.m. Thursday, August 30 stalls 3 through 12 beginning at 9:00 p.m. and finishing at 7:00 a.m. Friday August 31 stalls 38 through 46 beginning at 9:00 p.m.

10 and finishing at 7:00 a.m. or until the work is completed. Anyone planning on leaving or arriving after 9:00 p.m. on Thursday or Friday must plan on parking in the outside

12 parking lot as the entrance and exit areas will be blocked off by equipment while the garages are being cleaned. Also, all articles in the garage stalls must be removed,

14 including the bicycles. Any items left in the stalls will be considered trash and disposed of. Wall cabinets will be covered for protection. If you will be out of town during this

16 time period and plan to leave your vehicle in the garage, please notify us and provide access to car keys so that your car can be moved at the appropriate time.

18

    Thank you for your cooperation.

Appendix 2. Chris Applegate's Job Application Letter

(For use with Chapters 8, 10, and 12.)

419 North 10th Street
Star City, TN 37817
June 19, 1998

Express Rent-A-Car
255 East Melville Avenue
Knoxville, TN 37919

To Whom It May Concern:

An advertisement in the June 12, 1998 edition of the Knoxville <u>Journal</u> newspaper has informed me that Express Rent-A-Car is hiring entry-level managers for their expanding company. I believe my academic background and work experience have adequately prepared me to assume one of these positions.

In the Management program at the University of Tennessee, I received basic instruction on various topics introducing me to the many facets associated with successful management. Managerial responsibilities were discussed and significant consideration was given to the important dimension of manager-subordinate communication. I desire to use this knowledge to establish and/or maintain a productive and satisfying working environment within your organization.

In the Navy, I was the supervisor for the ship's electrical discrepancy repair team for about six people. I also performed supply functions for the electrical division and was fully responsible for the acquisition of parts and equipment as well as the generating and maintaining of all applicable financial reports. In addition, I worked part time all the way through college. These jobs have allowed me the opportunity to work and communicate with a wide variety of people from all over the country, a benefit which could prove to be very useful to a manager.

I would appreciate the opportunity to talk to you about how my training and experiences have prepared me for a management position at Express Rent-A-Car. Please feel free to call me at (423)555-1289 or write me to let me know a convenient time for you to talk to me.

Sincerely,

Chris Applegate

Chris Applegate

Appendix 3. Brian Carter's Job Application Letter

(For use with Chapters 1, 2, 12, and 13.)

The following letter was drafted in response to an ad asking "Recent Grads in Marketing" to apply for sales positions with "a national marketing firm of food products." A draft of the writer's résumé is given in Appendix 4.

83 Knight Avenue
Newark, NJ 27843

April 25, 1998

Mr. Hal Robinson
Bell Food Products
8100 Marshall Road
Camden, NJ 28193

Dear Mr. Robinson,

This June I will be graduating from Garden State College with a degree in Business Administration. I hope to follow a sales or marketing career. I am submitting an application for one of your sales positions advertised in the Sunday, April 23 edition of the Atlantic Times.

The courses we are required to take at Garden State for a sales/marketing degree reflect an in depth study of all aspects of this field. I am enclosing a résumé which further lists my educational background and work experience.

Also, I feel I am the type of person your company is looking for, hardworking, energetic, ambitious, and honest. My record of 50+ hours a week of summer employment reflects these qualities. I can offer your company an employee who is willing to give 110% all of the time.

If you feel I am the type of employee you are looking for please call me for an interview. I can be reached at 555-9813 any time convenient for you.

Sincerely

Brian Carter

Brian Carter

encl: résumé

Appendix 4. Brian Carter's Résumé

(For use with Chapters 1 and 7.)

Brian Carter

<u>PRESENT ADDRESS</u>   <u>After May 30, 1998</u>
83 Knight Avenue   7219 Whitman Drive
Newark, NJ   27843   Camden, NJ   28923
555-9813   555-7494

CAREER OBJECTIVES: MARKETING/MANAGEMENT FIELD

EDUCATION

                              Garden State College - Bachelor of
                              Business Administration '99

WORK EXPERIENCE

                              1997-1998 Cashier
                              Ingram's Gifts

                              1995-1997 Quality Control
                              County Seat D.C. Checked pullers orders
                              for errors, corrected errors, and filled
                              orders. Checked, priced and received
                              merchandise.
                              Also did inventory.

ACTIVITIES

                              B.A. Club
                              Intramural Softball
                              Tennis, rock climbing, skiing
                              Newark Affordable Housing Volunteer

Appendix 5. A Definition of Aphasia

(For use with Chapters 1, 3, and 11.)

The following draft was prepared as the introductory section of a pamphlet to be distributed by health professionals. The purpose of the pamphlet is to help explain aphasia to the family and friends of patients with this disorder. (Other sections will deal with prognosis and treatment.) Aphasia typically has a sudden onset and is most common in older persons.

"What is Aphasia?"

2

Aphasia is a language disorder caused by damage to brain tissue, usually
4   from a stroke, tumor, or trauma. The word <u>aphasia</u> comes from a Greek term
meaning "without speech." However, it is used to today to describe a disorder
6   of the higher-level mental processes involved in language, rather than to
describe lower-level disorders.

8

In most humans, the left hemisphere of the brain is believed to be
10   specialized for language; therefore, damage to certain areas of the left half of
the brain often results in aphasia. Two major types of aphasia can be defined,
12   according to the site of the damage and the syndromes that result. <u>Broca's</u>
<u>aphasia</u> (also known as <u>motor aphasia</u>) results from damage to the part of the
14   left hemisphere known as Broca's area. <u>Wernicke's aphasia</u> (also known as
<u>sensory aphasia</u>) results from damage to the part of the left hemisphere known
16   as Wernicke's area.

18

A patient with damage to Broca's area typically retains the use of content
words (nouns and verbs), but omits functors such as prepositions, articles, and
20   inflections. In addition, the Broca's aphasic typically speaks slowly, haltingly,
and without normal intonation. The patient's speech seems to require much
22   greater effort than that of a normal speaker.

24

The symptoms of Wernicke's aphasia are in many ways the exact opposite
of those associated with Broca's aphasia. In contrast with the slow, hesitant
26   speech of the Broca's aphasic, the speech of the Wernicke's aphasic is fluent--
that is, it is produced at a normal rate, with normal intonation, and with little
28   or no impairment of functors. The hallmark of Wernicke's aphasia is the
production of <u>neologisms</u> (new words).

Appendix 6. Memo Assessing an Employee's Writing

Huntwell Research Corporation, an engineering firm forced to lay off employees, offered to help them prepare letters of application for other positions.

PART A. (For use with Chapters 1, 2, 11, and 12.)

    An engineer asked for advice on the draft below of a letter to the Samson Corporation.

        I am interested in applying for the position of a software engineer.

2

        At Huntwell Research Corporation, I developed the QA department's formal software

4   using Mil-Std-2167 for their Global Positioning SW endeavor.

6   You will also notice that I also have 5 years of applicable experience in simulating fire-control, including missile simulations (red-eye, sparrow, etc.) in addition to being

8   Farringer's resident Tactical Systems Engineer.

10   If you peruse my resume it is self evident that I am an extremely competent individual in terms of my engineering capabilities: satellite systems, missile systems, communications,

12   industrial automation and so forth.

14   I am 60 years old--not obsolete by anyone's standards--mature with a proven track record, would rather work for 35K than $5.00 an hour as a security guard and in my personal

16   opinion "one heck of a systems design engineer"!

18   I would personally appreciate it if Samson Corporation does not see fit to offer me a position, if you would appraise my technical background and give me your opinion on why

20   I am not suitable for the position. If you will do this for me I will send you an autographed Xerox picture of the Minuteman Missile taken at Cape Canaveral in 1962; the

22   picture is a "classic" and appeared in Aviation Weekly.

PART B. (For use with Chapters 7, 11, 12, and 14.)

The personnel manager drafted the following assessment of the engineer's letter.

I have read and reviewed your letter. First of all even though you are writing to an
2   aerospace company does not mean the person reading your letter will understand all of the
initials that were used. Explain what the initials mean in the letter. Then, after you have
4   explained the initials, you can use the initials throughout the letter.

6   The first paragraph should contain where you heard about the job, the job you are applying
for, a statement about the letter, and something that will make the reader continue to read
8   the letter. Be convincing and let the reader know your interested in the job.

10   The body or main part of the letter should be about your qualifications and job
experiences. Your paragraphs two, three, and four should be the body of your letter but
12   expand them. Do not be egotistical about yourself or your job experiences. Tell your
reader more about what your Farringer's Tactile Systems Engineer job was. Should
14   include what the job required of you, was it hard, easy, were you in charge of the program
or did you just work on a team. All of this can give the reader a better understanding of
16   you and some idea whether you can work with others. This is very important for the
person who is doing the hiring because the job may warrant this type of person.
18
I would omit paragraphs five and six. The reader does not want to here about this kind
20   of stuff. The information in these paragraphs may just keep you from getting a job. Stay
away from this type of information in your letter.
22
Then a closing paragraph that contains a way for the reader to contact you, when you are
24   available for an interview, and when you can start work.

26   I would not give up your search for a job in the aerospace industry. You do need to
improve your letter and be more positive towards your reader.

Appendix 7. Clinical Guidelines for Electrosurgery

(For use with Chapters 8 and 10.)

The following is a draft of guidelines for use by veterinary students.

SELECTING THE ELECTRODE

The smallest electrode that will obtain the desired results should be used. Activated loop electrodes generate more energy during surgery than needle electrodes, resulting in a wider band of coagulation necrosis attributable to lateral heat production. Temperature increases in the adjacent tissue following use of the loop remain for longer periods of time than after use of a needle electrode. It has been calculated that a cooling interval of 15s is necessary to properly dissipate heat between successive entries with a loop electrode. Lateral heat production adjacent to a fine wire needle electrode requires a cooling period of 8s.

SPEED OF CUTTING

To achieve the best result, the cutting should be at a brisk speed. The motion should be at a rate that incises the tissue adequately without a cooked or charred appearance. The longer the electrode is in contact with the tissue, the greater the amount of lateral heat generated. It has been concluded that moving the electrode through tissue at a rate of 7mm/s was compatible with good clinical technique and minimal production of lateral heat.

POWER LEVEL

The power level is slowly increased until a satisfactory speed of cutting is achieved without resistance. At too low power levels there will be either failure to cut or resistance to cutting. There will be exaggerated arcing if the power is too high. Sparking also occurs if the tissues are dry, if the passive or ground electrode is not used, operating in diseased tissue or using a defective electrosurgical unit. If a malfunctioning unit is suspected, testing and calibration of electrosurgical devices has been made more accurate by the development of programmable waveshape generators and power output meters. These measurements are made to insure that a device is operating at a level sufficient to perform electrosurgical techniques with low radio-frequency leakage current to avoid the risk of burns to the patient or disruption of other electromedical equipment. This instrumentation also allows better monitoring and control to conduct experiments evaluating the effects of electrosurgery on tissues.

Appendix 8. Research Proposal on "Erosion at Sawyer Road Park"

(For use with Chapters 1, 2, 3, 4, and 18.)

The following draft of a research proposal was prepared by a student for a technical writing instructor. It describes a longer report to be completed as one of the course requirements.

Project Summary

2

    Observation by employees at Belleville's Parks and
4   Recreation Department revealed that the city's new one
    million dollar baseball facility is eroding at an alarming
6   rate. I propose that the Recreation Board commission a
    study to determine the feasibility of repairing the eroding
8   baseball fields. This study would cover all aspects of the
    project, such as repairing damaged fields, installing a new
10 drainage system, grading fields and surrounding areas, and
    preventive measures to take while caring for the fields.
12

Project Description

14

    Since the inception of Belleville's Parks and Recreation
16 Department's Sawyer Road Project in 1986, there has been a
    need to observe problems associated with landfills. In
18 1986, the Sawyer Road Wastewater plant was closed and made
    into a landfill. Sawyer Road Park was then constructed on
20 the site in the next two years. As it has come to the
    attention of many park-goers, the baseball fields have had
22 terrible erosion problems. I propose that the Recreation
    Board commission me to carry out a study to evaluate the
24 erosion problems in the park. I will make investigations
    into equipment and methods needed to correct the problem and
26 prevent its occurrence. Completion of this project would
    ensure the citizens of Belleville a vital and important
28 recreation facility. The importance of these baseball
    facilities to the youth of Belleville is priceless. The
30 fields serve as a medium where over 2000 boys and girls
    interact and participate in organized sports. If these
32 fields are left to erode at the present rate, the sinkholes
    that develop will threaten the safety of the children and
34 eliminate summer baseball and softball leagues.

36   Erosion is common in Western Kentucky, and this landfill
    is typical of many Western Kentucky fields. Summer showers
38 wash away sediment, and if left untreated can ruin a
    beautiful and expensive park. If action is taken in the
40 near future, the correction of erosion costs should be
    moderate, but if not, the cost would eventually surpass the
42 initial cost of the facility. Immediate action would halt
    any erosion, and corrective measures would repair any damage
44 done. Correction of erosion projects has been carried out

successfully by many farmers in the Western Kentucky area. Area farmers can attest to the feasibility of a correction project. The land is their lives and care of the land is their specialty. The equipment and knowledge are available that would correct all erosion problems found in the Sawyer Road fields. Should the benefits outweigh the cost and problems created by a correction project, it would seem feasible.

Plan of Work

The proposed study breaks into three steps:

1. Conducting research into the location and extent of all erosion present in Sawyer Road Park.

2. Finding out how much equipment and what methods should be used for repair and prevention of erosion.

3. Analyzing the collected data and writing the report.

I will locate all areas of erosion and make a thorough check of the surrounding park areas. This check would help locate any future trouble spots that may crop up in the future. For instance, I will seek the advice of several area farmers on the degree of erosion and whether or not treatment is needed.

For the equipment, I will correspond with area construction companies to obtain estimates for the cost of equipment and labor. To obtain favorable methods for the repair and correction of erosion, I will contact several area farmers who have dealt with the problem on their own farms. Also, I will contact officials at Kentucky's Department of Transportation and Belleville Public Works to find out what methods were successful for them.

Personal Qualifications

I am a junior majoring in Civil Engineering. I am extremely familiar with Sawyer Road Park, having worked in the area for four years on weekends and during summers. I have seen the progression of the erosion over the past couple of years.

Appendix 9. Executive Summary on "ProWear Shoes"

(For use with Chapters 4 and 6.)

The following draft is of an executive summary to accompany a report on "Increasing ProWear Shoes' Appeal to Older Americans."

The idea that "America is Aging" is now a reality and not just a
2   commonly voiced view. ProWear has been concentrating its advertising
efforts towards younger, more affluent Americans (15-32) and ignoring the
4   larger, older American market. By the year 2020, the number of older
Americans will increase by 74 percent; the number of Americans less than
6   50 years of age will increase only 1 percent. ProWear could increase its
profitability by advertising current products to the older market. A product
8   that offers quality, arch and ankle support, a lot of comfort, and a
reasonable price is what these consumers wish to have. ProWear's current
10 product line offers all of these requirements as well as a stylish appeal.
ProWear could target the older market with few modifications to its product
12 line. Many advertising attempts by other companies have only succeeded in
offending older Americans, and ProWear needs to create an advertising
14 campaign that will prevent this. ProWear should follow two general
guidelines in their advertising for this market. No one will buy a product
16 that portrays them as grotesque, mentally unstable, or silly, so ProWear
must avoid any age stereotyping. ProWear should use advertising that
18 takes into account perceptual and cognitive changes that accompany age:
1) changes in vision, 2) changes in hearing, 3) changes in motor skills, 4)
20 changes in condition, and 5) psychological factors. Demographic changes
are presenting ProWear with the opportunity to expand sales and
22 profitability by targeting older Americans. If ProWear follows the advertising
guidelines recommended above, they will succeed in capturing this growing
and increasingly active market.

Appendix 10. Report on "Improving Employee Training at Becker Foods"

(For use with Chapters 1, 2, 4, 8, 11, 13, 14, and 16.)

The following is a draft of a recommendation report prepared for the management of Becker Foods.

EXECUTIVE SUMMARY

2   While Becker Foods' sales are booming, there is a need for consistently skilled employees. Techniques to overcome this problem is addressed.

4   The strategies chosen to train and increase the skill of the cashiers, especially, are: formulating a structured training program, appointing a specific trainer and positioning a

6 simulation machine in the back. Each alternative was addressed and the formulation of a structured training program and specifying a specific trainer was the most efficient

8 alternatives for Becker Foods.

10   A drawback to the training program concerns training the cashiers during non-peak hours for the opportunity to efficiently train, which consists of early morning or late

12 evening hours.

14   In support of a structured training program, the current program used by another company, Melrose Foods, has been mentioned. Their structured five-day program is

16 analyzed and ways of adapting for Becker is considered.

18   A drawback to the simulation machine includes the initial investment to purchase the machine.

20

      It is thus concluded that Becker Foods should design a structured training program

22 and designate a specific trainer to perform the program.

24 ### INTRODUCTION

26   While Becker Foods' sales are booming, there is an abundance of grocery stores within the Southville area. Although, they have shown through their current clientele and

28 profits that they are prosperous. Due to the competition, Becker needs to form their own market niche in order to "shine" above the rest.

30

      Currently, they have a reputation for excellent customer service and satisfaction.

32 Due to this reliance on strong customer service, their is a need for skilled workers.

34   The first portion of this report will analyze the current training program and policies in effect and the training program used by another store in the industry. The second

36 portion will analyze alternative solutions to their training situation to better serve the customer. Finally, provide recommendations for actions to be taken by the company to

38 improve their training program.

In collecting data for this report, personal employment experience of the author with this company and another company in the grocery industry, will be used. The sequence of this report will be to address the problem, propose a solution, determine the feasibility of the solution.

This analysis leads to the recommendation to formulate a structured training program and an administrator of training to provide a concise form of training for the employees.

BRIEF OVERVIEW

Becker Foods relies heavily on customer satisfaction and friendly employees. If you question any of the customers they will explain to you that they appreciate the attention that is given to them by the employees and especially upper management. With only 85 employees in the store, there is a sense of family that the customer sees, senses and enjoys being made a part of.

The grocery market can be especially profitable if operated efficiently and in the correct context.

Problem

Becker Foods has an overall problem with their training program. There is no structured program or designated trainer. The employees, mainly cashiers, are trained by current cashiers and are instantly confronted with customers. The customers seem to be rather understanding but it is rather unnerving to the cashier trying to learn. Currently, cashiers train for one day and with their next shift, they are on their own relying on the help of near-by cashiers. Cashiers have different styles and understandings of the register and without a consistent trainer there could be some missed information or misunderstood information for the new employees.

Melrose Foods' Different Perspective

Currently, Melrose Foods administers a five-day training program. According to Leslie Morgan, Customer Service Manager of Melrose Foods, they administer this training program in groups of 3-5 employees. They have the employees do a series of training rules to get the hang of the register and to gain an understanding of the different departments. All is done on the registers without customer interruptions. Ms. Morgan states that, "Our employees are confident when they confront the customer for the first time. Granted they are nervous, but they (the cashiers) feel they know the register well enough so they can overcome their fears easily."

Before the cashiers are through with their training program, they have been quizzed on their produce codes, different department categories, the register as a whole, and situational analysis.

Recommendations

Becker Foods should adopt a structured program, such as the one that Melrose Foods enforces. Becker needs to appoint a specific trainer and program so that all is covered and consistently covered.

The five-day program is a little extreme for Becker to enforce, therefore, a program that consisted of eight hours or two days would be sufficient for the cashier to learn the needed material before confronting customers.

The management would need to set up a sufficient time that it would be slow enough so as to close a register and use it only for training.

I believe that they should continue to train only one cashier at a time due to the limited number of registers.

The drawback to this situation is that the non-peak times occur either late at night or during the morning hours and with many of the employees being college students the morning hours are used for school.

Another recommendation would consist of setting up a simulator register in the back. So that the cashier could be trained without consistent interruptions and could maintain a training atmosphere. This would also involve an appointed trainer. This would also eliminate the need to close a register and could adhere to the convenience of the trainer or the new employee.

This may be an initial investment to hire the trainer--which could be a part-time trainer and when not training they could cashier to cut back on the initial expense--and to purchase the simulator machine, but in the long run I believe it will pay off with increased consistency and overall efficient performance.

CONCLUSION

The solution which appears to be the most effective should be implemented in order to maintain the consistent quality customer service that Becker is known for and to become more efficient involving their training.

Evaluations should be conducted periodically to determine effectiveness. If the programs appear to not perform effectively, they need to be modified or discontinued and new solutions adhered to.

Becker Foods possesses the expertise to perform the tasks required for efficient training. The structured training program can be effectively initiated through appropriate measures. By structuring the training program, efficiency and overall satisfaction by the employees and customers will be achieved.

Appendix 11. Report on "Security Methods for Ashley's Clothing Store"

(For use with Chapters 1, 2, 4, 6, and 18.)

The following is a draft of a report prepared for the Manager of Ashley's Clothing Store.

WHAT SECURITY METHOD SHOULD ASHLEY'S ADOPT?

INTRODUCTION

2

    Ashley's at Kenwood Mall is one of four clothing stores
4 owned and operated by Ruston, Inc., of Philadelphia.
Ashley's is a retail clothes store that caters to teenage
6 girls.

8   Over the past five years Ashley's has experienced a
marked increase in theft, or what is known in retail as
10 inventory shrinkage. This is evident through the increasing
loss the store has experienced every time the yearly
12 inventory report comes out. Also, it can be seen daily in
the store in the form of empty hangers found at the end of
14 the day. Not only that, but there have also been incidents
of customers "exchanging" their clothes in the fitting
16 rooms. This is when a customer walks into a fitting room to
try on an outfit and will walk out with that outfit on
18 underneath their coat, leaving their clothes behind on the
hangers.
20

    Ashley's currently has no security policy in
22 implementation. They depend solely on the employees to be
aware of any potential, or ongoing shoplifting within the
24 store. The employees are not formally trained in any way,
but are told to trust their own judgment.
26

    The purpose of this study is to recommend a feasible
28 solution to help control the inventory shrinkage that
Ashley's has been experiencing. The report will look at
30 various security methods available and make recommendations
of programs that should be implemented in order to reduce
32 theft in the store. The report is structured in such a way
that the first portion will include an analysis of different
34 methods of security currently available on the market. This
will be followed by my recommendation, and finally a
36 conclusion.

38   Four different methods of security were looked at to
determine which would be best to implement at Ashley's.
40 These include employee awareness, Colortags, Electronic
Surveillance System, and the TellTag system. It is my
42 recommendation that both employee awareness and the
Electronic Article Surveillance system be put into effect

2   immediately. The benefits of the programs will be discussed
     further into this report.

4

                    METHODS TO DETER SHOPLIFTING

6

8   Both the advantages and the disadvantages were pointed
     out in evaluating each method. Some of the factors that
     were considered include price, effectiveness, and ease of
10   implementation.

12   **A. Employee Involvement**

14   This would range from training employees on security
     awareness, to preventive steps that sales associates could
16   take to reduce theft. According to Sandy Katz, director of
     loss prevention for General Mills, "Awareness education is
18   the biggest key to improving inventory shortage" (Schulz,
     87). This can be done through workshops offered to the mall
20   employees, and also through in-house video tapes. These
     tapes would point out suspicious behavior to look for and
22   actions associates can take to prevent a potential
     shoplifter. There are also preventive measures that could
24   be taken which don't require extensive training. One
     example would be to start locking the fitting room doors.
26   This would require all customers to be escorted into the
     fitting rooms by a sales associate, thereby monitoring what
28   people bring in and out of the fitting rooms. Another
     security measure would be the layout of the store
30   merchandise. Smaller items, such as jewelry, could be
     placed near the wrap desk where the employees could keep a
32   better eye on them. High ticket items could be placed away
     from the door, making them not as easy of a target for a
34   quick getaway.

36   Employee awareness would be the least expensive route
     for the store to take in reducing inventory shrinkage.
38   Also, Shelly Connors, regional loss prevention manager for
     Best Products feels that "Security-conscious employees are
40   the most effective deterrent against inventory shortage"
     (Employee Involvement, 80). The program would be very easy
42   to implement, for all it involves is heightening employee
     awareness.
44
         But employees can't be everywhere at once. When the
46   store is busy it is still difficult to monitor all the
     customers in the store.
48

     B. Colortags
50

         The method of Colortag relies on dye capsules implanted
52   inside the garment tags. If the Colortag is removed
     improperly, the capsule ruptures, spilling a permanent stain
54   dye all over the garment, making it worthless (Schulz, 90).

2   The sales associates would be in charge of properly removing the tags at the point of sale (POS).

4   The Colortag method is very cost efficient. Compared to other methods available, the cost is minimal. Whether the
6   "thief" wanted to resell merchandise of keep it for themselves, the removal of the tag would make the garment
8   unwearable. This would take away the motive for theft (Schulz, 90). Because no electronics are involved, the
10   system would be easy to implement in the store.

12   However, if a sales associate forgets to remove a tag at the point of sale, an honest customer may suffer. If the
14   customer brings home a garment and tries to remove a Colortag, they may not only end up with permanent dye all
16   over the garment, but also all over their home. Also, because the store caters to younger people, there might be
18   acts of vandalism once the system becomes known. Another problem is that accidental damage by both employees and
20   customers could be a frequent result when dealing with permanent dye. Also, the Colortag system is fairly new on
22   the market; therefore, it has not been proven to be effective.
24
     C. Electronic Article Surveillance
26
     The most prevalent method of EAS involves placing a
28   small circuit, or target on a garment. The most common form of these would be the large, white, plastic tags which can
30   be found on garments in retail stores all over the country. If the target passes through a field of a specific
32   frequency, the target is activated and a signal is emitted that can be read by a detection device. The target itself
34   contains no power source, it is totally passive (Radio Frequency, 11). The detection device is usually in the form
36   of two pedestals standing on either side of the entrance of the store. This system would require the sales associate to
38   remove the tag after the merchandise has been purchased.

40   The tag is read without requiring physical contact or line-of-sight access to the detection device. The tag needs
42   only to be in the proximity of the device (Radio Frequency, 11). This method would not require constant surveillance
44   of the customers by the employees. Because the tags are reusable, the store would only have to cover the initial
46   cost. The maintenance would then be paid for through the reduction of costs in inventory shrinkage (Radio Frequency,
48   11).

50   A problem is that the large tags on garments may inhibit shopping. Customers may be less apt to try something on
52   that has a big plastic tag on it (The Reality, 93). A sales associate may forget to remove a tag, which may result in
54   embarrassing an honest customer (Smarter Tags, 295). The

2   initial cost of the EAS system would make it somewhat
    expensive to implement. And the alarm will not completely
    eliminate shoplifting. Some people won't be inhibited by
4   the sound of an alarm, and may just take the merchandise and
    run. When in the privacy of the fitting rooms, customers
6   may try to remove the tags by force (Smarter Tags, 295).

8   **D. TellTags**

10   The TellTag system is an advanced form of the Electronic
    Surveillance System. It works with the same concept of
12 frequency tags placed on the merchandise. It differs in the
    fact that it alarms directly from the tag if merchandise is
14 taken from the store. The TellTag is triggered not by the
    bulky pedestals of typical EAS systems, but rather by a wire
16 mesh antenna which can be easily hidden in a sign, a picture
    frame, or the floor (Smarter Tags, 295). Not only that, but
18 if the tag is tampered with in any way in the fitting rooms,
    an acoustic tone detector amplifies the sound of the tag
20 going off, which will alert a sales associate. Also, the
    tag reminds the employee to remove it by beeping four times
22 when it comes into contact with the equipment at Point of
    Sale (POS) (Abend, 63).
24
        Advantages are that the TellTag system will help to
26 alleviate theft in the fitting rooms. Because a reminder
    alarm is sounded, this will eliminate embarrassing a
28 customer who has made an honest purchase. Like the typical
    EAS system, the sales associate can concentrate more on
30 selling, rather than monitoring all the customers in the
    store.
32
        On the minus side, the cost of the TellTag system would
34 be four to five times higher than the typical EAS system
    (Abend, 63). Also, the tags on the garments would be the
36 big, bulky, plastic tags, like the EAS system. Accidental
    tampering with the tags by either the employees or the
38 customers would trigger a false alarm. Also, like the EAS
    system, there are people who are simply undeterred by the
40 sound of an alarm and will continue to steal merchandise
    (Smarter Tags, 295).
42
    E. Recommendations
44
        After weighing the pros and cons of the available
46 methods, I feel that by implementing both employee awareness
    and the Electronic Surveillance System (EAS), Ashley's will
48 be successful in reducing theft. This strategy would enable
    the store to have a fairly sophisticated surveillance
50 system, that would be backed up by trained, knowledgeable
    employees. All this can be done at a fairly reasonable cost
52 that will soon pay for itself in the way of reducing
    inventory losses.
54

CONCLUSION

2

4
    Ashley's would benefit a great deal if it were to implement and integrate these security methods successfully. By backing up a new security system with increased employee

6
awareness, the inventory shrinkage problem can be a thing of the past.

REFERENCES

Abend, Julie. "More Deterrents to Theft." <u>Stores</u>, June 1989: 61-63.

"Employee Involvement is Key to Store Security." <u>Chain Store Age Executive</u>, April 1989, 80.

"More Deterrents to Theft." <u>Stores</u>, June 1989: 61.

"Radio Frequency Identification." <u>Stores</u>, Sept. 1989: 11-12.

"The Reality of EAS: It's in the Tag." <u>Chain Store Age Executive</u>, July 1989: 93.

Schulz, David. "Specialty Problems." <u>Stores</u>, June 1988: 87-90.

"Smarter Tags Combat Smarter Thieves." <u>Chain Store Age Executive</u>, May 1990: 295.

Appendix 12. A Discussion of Prenatal Exposure to Cocaine

(For use with Chapters 4, 6, 7, and 18.)

The following draft is a section from an article on the effects of prenatal exposure to cocaine. The article was intended to inform speech-language pathologists and educators who might encounter children who had been exposed prenatally to cocaine.

Physical Effects of Prenatal Exposure to Cocaine

2   The physical effects that result from prenatal exposure to cocaine can be categorized into four categories: prematurity, intrauterine growth retardation, malformations, and
4   consequences of vasoconstriction. "Lack of parental care and poor prenatal nutrition together often lead to low birth weight and/or significant prematurity, factors that
6   further reduce the infant's chances for optimal development" (Griffith, 1992, p. 31). "A developmentally immature fetus is medically at risk for a number of reasons.
8   Outcomes related to underdeveloped livers such as jaundice, and to underdeveloped lungs such as apnea, anoxia, and meconium aspiration may all result in damage to the
10   under developed central nervous system" (Williams & Howard, 1993, p. 67). Intrauterine growth retardation is indicated by lower birth weight, shorter length, and a
12   smaller head circumference. Cocaine can be directly attributed to only 25 percent of the weight reduction (Phelps & Cox, 1993). The weight and height differences of the
14   children prenatally exposed to cocaine "caught up with those of nonexposed children" (Shriver & Piersel, 1994, p. 168) at the 1-year age level. "Head circumference is the
16   best single indicator of normal brain growth and development. A small head size at birth that fails to catch up is a significant predictor of poor development" (Griffith,
18   1992, p. 32). Malformations that can be attributed to cocaine include cardiac, central nervous system, and gastrointestinal anomalies most commonly displayed as prune
20   belly. However, the majority of problems among infants with cocaine exposure are the effects related to vasoconstriction. The cocaine causes the constriction of the blood
22   vessels to the placenta. "The sudden and intense blood vessel constriction can cause the placenta, which is primarily composed of veins, to tear away prematurely from the wall
24   of the uterus" (School Safety, 1992, p. 17). Vasoconstriction may also result in fetal anoxia (lack of oxygen), which can lead to spontaneous abortion, cerebral infarction
26   (stroke), and irregular heart rate (School Safety, 1992; Williams & Howard, 1993).

References

Chapter 2. Revising Persuasive Prose

Toulmin, S., *The Uses of Argument* (Cambridge: Cambridge UP, 1985).

Chapter 3. Revising Graphics

Manning., A., "The semantics of technical graphics," *J. of Technical Writing and Communication*, 19 (1989), 31-51.

Chapter 5. Revising for Cohesion

Campbell, K. S., Theoretical and Pedagogical Applications of Discourse Analysis to Professional Writing (Louisiana State University, Ph.D. dissertation, 1990).

Campbell, K. S., "Structural Cohesion in Technical Texts," *J. of Technical Writing and Communication*, 21 (1991), 221-237.

Chapter 9. Revising for Parallel Structure

Parker, F. *Linguistics for Non-Linguists* (Boston: Little, Brown, 1986; Allyn & Bacon, 1991), pp. 71-73.

Chapter 10. Revising Active and Passive Voice

Ferguson, K. S., and Parker, F., "Grammar and technical writing," *J. of Technical Writing and Communication*, 20 (1990), 357-367.

Kies, D., "Some stylistic features of business and technical writing: The functions of passive voice, nominalization, and agency," *J. of Technical Writing and Communication*, 15 (1985), 299-308.

Riley, K., "Passive voice and rhetorical role in scientific writing," *J. of Technical Writing and Communication*, 21 (1991), 239-257.

Chapter 11. Word Choice

Parker, F., and Riley, K., *Exercises in Linguistics* (Boston: Little, Brown, 1990; Allyn & Bacon, 1991), pp. 40-41.

Chapter 12. Revising to Improve Your Tone

Brown, P., and Levinson, S., "Universals in language usage: Politeness phenomena," in E. N. Goody (ed.), *Questions and Politeness: Strategies in Social Interaction* (Cambridge: Cambridge UP, 1978), pp. 56-310.

Campbell, K. S., "Explanations in negative messages: More insights from speech act theory," *J. of Business Communication*, 27 (1990), 357-375.

Campbell, K. S., Riley, K., and Parker, F., "*You*-perspective: Insights from speech act theory," *J. of Technical Writing and Communication*, 20 (1990), 189-199.

Riley, K., "Speech act theory and degrees of directness in professional writing," *The Technical Writing Teacher*, 15 (1988), 1-29.

Riley, K., and Parker, F., "Tone as a function of presupposition in technical and business writing," *J. of Technical Writing and Communication*, 18 (1988), 325-343.

Chapter 13. Editing Punctuation

Meyer, C. F. *A Linguistic Study of American Punctuation* (New York: Peter Lang, 1987).

Chapter 17. Editing Pronoun Reference

Parker, F. *Linguistics for Non-Linguists* (Boston: Little, Brown, 1986; Allyn & Bacon, 1991), pp. 5-9.

Chapter 18. Editing References to Other Sources

American Psychological Association. *Publication Manual of the American Psychological Association* (revised edition). Lancaster, PA: Author, 1994.

Dodd, J. S. (ed.) *The ACS Style Guide: A Manual for Authors and Editors*. American Chemical Society, 1997.

Gibaldi, J. and Achtert, W. S. *The MLA Handbook for Writers of Research Papers* (4th edition). New York: Modern Language Association, 1995.

Huth, E. J. *Scientific Style and Format: The Council of Biology Editors Manual for Authors, Editors, and Publishers* (6th revised edition). Cambridge University Press, 1994.

The Chicago Manual of Style: The Essential Guide for Writers, Editors, and Publishers (14th edition). University of Chicago Press, 1993.